Springer Tracts on Transportation and Traffic

Volume 17

About this Series

The book series "Springer Tracts on Transportation and Traffic" (STTT) publishes current and historical insights and new developments in the fields of Transportation and Traffic research. The intent is to cover all the technical contents, applications, and multidisciplinary aspects of Transportation and Traffic, as well as the methodologies behind them. The objective of the book series is to publish monographs, handbooks, selected contributions from specialized conferences and workshops, and textbooks, rapidly and informally but with a high quality. The STTT book series is intended to cover both the state-of-the-art and recent developments, hence leading to deeper insight and understanding in Transportation and Traffic Engineering. The series provides valuable references for researchers, engineering practitioners, graduate students and communicates new findings to a large interdisciplinary audience.

Indexed by SCOPUS, WTI Frankfurt eG, zbMATH, SCImago.

More information about this series at http://www.springer.com/series/11059

Polat Gülkan · Alp Caner ·
Nurdan Memisoglu Apaydin
Editors

Developments in International Bridge Engineering

Selected Papers from Istanbul Bridge Conference 2018

Springer

Editors
Polat Gülkan
Baskent University
Ankara, Turkey

Alp Caner
Civil Engineering Department
Middle East Technical University
Ankara, Turkey

Nurdan Memisoglu Apaydin
Ministry of Transport & Infrastructure
DLH Marmaray Bolge Mudurlugu
Uskudar Istanbul, Turkey

ISSN 2194-8119 ISSN 2194-8127 (electronic)
Springer Tracts on Transportation and Traffic
ISBN 978-3-030-59168-7 ISBN 978-3-030-59169-4 (eBook)
https://doi.org/10.1007/978-3-030-59169-4

This Springer imprint is published by the registered company Springer Nature Switzerland AG
The registered company address is: Gewerbestrasse 11, 6330 Cham, Switzerland

Preface

Bridge engineering is a unique branch of structural engineering. It occupies an eminent standing in the broad spectrum among disciplines of applied technology because it straddles into architecture, aesthetics and material science in meeting new challenges that confront urban societies in ensuring that the movement of people, goods and services between population and industrial centers proceeds safely and rapidly with no interruptions. This objective can be met only with robust teamwork among professionals who play different roles in the constructing of an object that has merit of societal utility. Facilitation of exchange of experience and knowledge among these actors represents a worthwhile objective for many professional associations.

The Third İstanbul Bridge Conference was hosted during November 5–6, 2018, by the Turkish Association of Bridge and Structural Engineering, the national branch for IABSE, as were its two precedents. That such a conference was held in İstanbul, a sprawling metropolis that straddles two continents separated by a strait is emblematic of the global significance of modern transportation systems for increasingly mobile societies. The Strait of İstanbul, currently spanned by three suspension bridges and two subaqueous crossings, will soon be traversed by two other major tunnels, one intended for rail and the other for road vehicles. Thus, bodies of water no longer represent barriers that transportation networks must scale in the interest of societal economic and social development, only opportunities and challenges for engineering in meeting the need for improved transportation systems.

Starting from 2014, the Turkish Association has organized the biennial İstanbul Bridge Conferences that have attracted global interest. Similar to its two antecedents, iBridge 2018 was organized to provide a continued exchange platform to develop safe, sustainable and economically feasible alternatives for bridge research, design and construction worldwide. The conference was attended by about two hundred delegates who represented the design and construction industry, academia and the regulating authorities. During two days, it facilitated a broad survey of many facets of bridge engineering, from the exalted 2023-m-long span of the suspension structure currently under construction across the Çanakkale (Dardanelles) Strait to the unexpected and tragic collapse of the Morandi Bridge in

Genoa that disrupted the freeway passage through the city, causing hardship to the national economy in Italy. While the first invokes the glory of building a record-breaking structure scheduled for completion in 2023, that elegantly rises above the historic shores of the fabled waterway, the latter is a reminder that durability and health assessment of major bridges will gain increasing importance as our current stock of bridges age, and can no longer be considered automatically as safe for current loads and traffic. Engineers must find clever solutions to assess the condition of bridges, using technologically innovative ways that ensure that assessments are correct and rapidly reached. One invited paper in the book is a comprehensive survey of the state of the art in this important area.

This book contains a selection of presentations at the conference. Its principal parts have been arrayed to cover broadly the following headings:

- Invited Papers (3 contributions)
- Bridge Design (8 contributions)
- Bridge Construction (8 contributions)
- Bridge Extreme Event Loads: Earthquake, Wind and Fire (3 contributions)

This text has been arranged as a readily accessible resource for the practicing engineer who constantly needs to bring innovative solutions to emerging problems in conception, design, construction and maintenance of bridges. Protection from extreme natural events, harmful effects that threaten bridge durability and innovative technologies for continuous monitoring of bridge behavior are among its highlights. The Editors hope that it will be a useful reference to every bridge engineer who wishes to take advantage of the experience of the community of bridge professionals.

The Third İstanbul Bridge Conference had many supporters and sponsors. These included the Ministry of Transportation and Infrastructure, General Directorate of State Highways, General Directorate of State Railways, Turkish Contractors Association, Turkish Society of Structural Steel, Galvanized Steel Manufacturers Association, the Turkish chapter for International Association for Bridge Maintenance and Safety (IABMAS) and the Bridge Engineering Association. The Turkish National Group for IABSE is pleased to express its gratitude for the support they have provided toward the success of the conference.

Finally, the Editors wish to put on record their gratitude to Ms. Banu Akman, who served as the designated Conference Secretary, effectively carrying out correspondence duties among the authors, editors and the publisher, cajoling sponsors and coaxing tardy authors to meet deadlines. But for her efficient and cheerful contribution, the book would not have arrived at its happy end.

Polat Gülkan
Alp Caner
Nurdan Memisoglu Apaydın

Contents

Invited Papers

Numerical Investigations on the Collapse of the Morandi Bridge

G. M. Calvi, M. Moratti, N. Scattarreggia, V. Özsaraç, P. M. Calvi, and R. Pinho

Abstract The collapse of a relevant portion of the Polcevera river viaduct, located in Genoa (Italy) and also known as Morandi Bridge, is herein object of a numerical investigation. The bridge was designed in the early 1960s by Riccardo Morandi, a well-known Italian engineer, and opened to the public in 1967. The collapsed part of the bridge essentially comprised an individual self-standing structure spanning 171 m and two simply-supported connecting Gerber beam systems, each spanning 36 m from the self-standing structure to the adjacent portions of the bridge. A previous investigation, by Calvi et al. (Calvi et al. in Struct Eng Int 29:198–217, 2019), indicated that the collapse of the bridge could have potentially been triggered by a sudden loss of effectiveness of one stay (e.g. due to the failure of either the deck-stay or the antenna-stay connections), which in turn would have induced a flexural–torsional-shear failure of the bridge deck. In the current work, therefore, the attention is turned into a more detailed study of the latter, through the employment of detailed Finite

Disclaimer: The authors emphasize that the investigations presented in this paper have been completed in December 2018 and thus reflect the state of knowledge of that time. At the time of publication of this book, in May 2021, relevant additional evidence had been rendered public, including a video of the collapse that enlightens some aspects of the collapse sequence. None of the information or evidence that has become available after December 2018 has been considered in this work.

G. M. Calvi (✉) · M. Moratti
Studio Calvi Ltd, Pavia, Italy
e-mail: gm.calvi@eucentre.it

R. Pinho
Mosayk Ltd, Pavia, Italy

G. M. Calvi · N. Scattarreggia · V. Özsaraç
IUSS, Pavia, Italy

R. Pinho
University of Pavia, Pavia, Italy

P. M. Calvi
University of Washington, Seattle, USA

G. M. Calvi
Eucentre Foundation, Pavia, Italy

P. Gülkan et al. (eds.), *Developments in International Bridge Engineering*, Springer Tracts on Transportation and Traffic 17, https://doi.org/10.1007/978-3-030-59169-4_1

Element Analysis modelling, with a view to corroborate, or not, the results of the aforementioned previous study. As in the latter, the development of such model was based entirely on publicly available material that has so far been rendered online accessible. It is shown that should the support provided by one of the bridge stays be removed, the ensuing flexural–torsional-shear demand on the deck would be such that the latter would be led to the type of collapse observed in the field.

1 The Morandi Bridge

The original design of the entire viaduct is described in Morandi [7, 8], being then revisited, in particular for what concerns its collapsed part, in the work of Calvi et al. [3]. Herein, only a brief overview of the viaduct's structural configuration and main components is therefore given, with the exception of the main deck, which, being the main object of study in this work, is necessarily described more in-depth (based on the drawings that are currently available online).

1.1 Structural Configuration

The entire viaduct structure is shown in Fig. 1, with each of its twelve support points being numbered sequentially from the Sava end, with piers 9, 10 and 11 constituting the so-called "balanced systems" (Fig. 2); it was pier number 9 that collapsed on the 14th of August 2018.

Each balanced system comprises the following main elements:

a) A *pier* with eight inclined struts (with cross-section varying between 4.5 × 1.2 to 2.0 × 1.2 m) that props the deck over a distance of about 42 m.

b) An *antenna* with two A-shaped structures (element cross-section varying between 4.5 × 0.9 to 2.0 × 3.0 m) that converge about 45 m above the deck level.

c) A *main deck* with a five-sector box section of depth variable between 4.5 and 1.8 m, an upper and lower slab 16 cm thick, and six deep webs with thickness varying between 18 and 30 cm. In its final configuration, the deck of the *balanced*

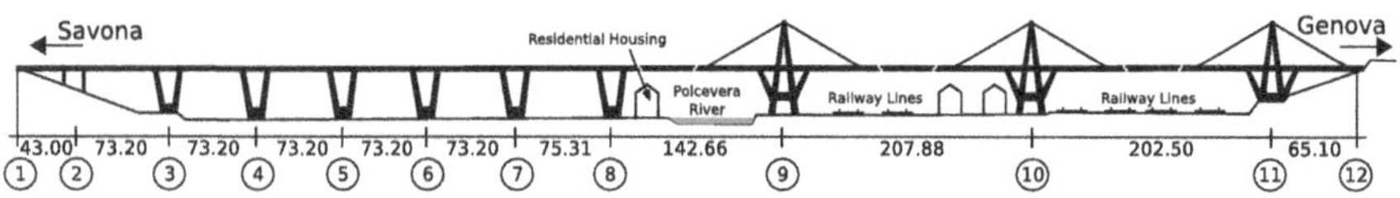

Fig. 1 Schematic of the piers and distances between each support of the viaduct, with the three balanced systems shown to pass over residential areas, numerous transportation lines and the Polcevera river (although not shown, the area between piers 1 and 8 is also heavily industrialised)

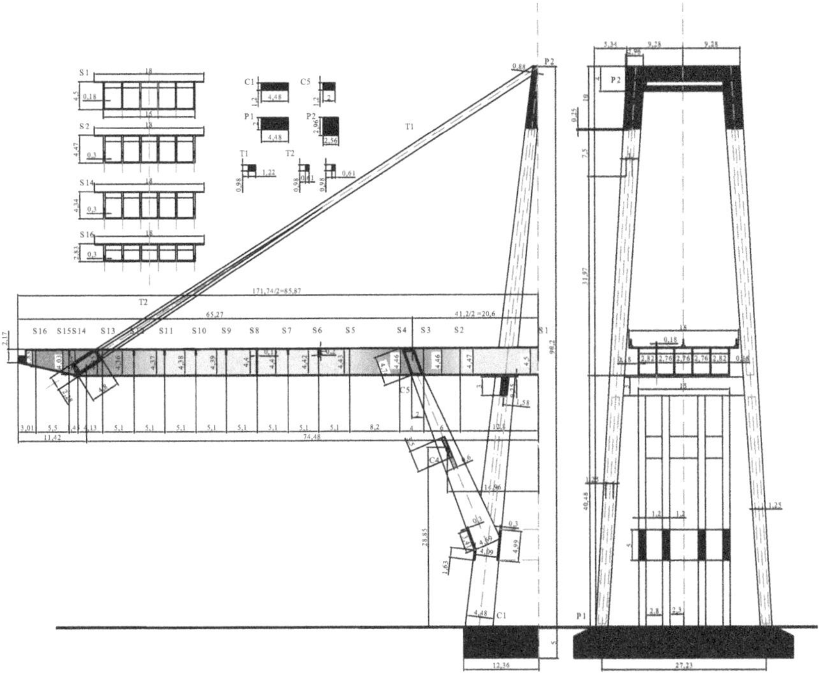

Fig. 2 Longitudinal and transversal section of one of the "balanced systems" that constituted the long-span portions of the viaduct

system 9 was 172 m long and supported at four locations: two of these from underneath the deck, provided by the pier inclined struts at the aforementioned spacing of 42 m, and the other two from above, provided by the cable stays at a distance of 152 m. There was therefore no connection between the deck and the antenna. Two 10 m cantilevers completed the deck length.

d) Four *transverse link girders*, connecting stays and pier trusses to the deck.
e) Four *cable stays*, hanging from the antenna's top and intersecting the deck at an angle of about 30°.
f) Two *simply-supported Gerber beam spans* connecting the balanced system to the adjacent parts of the bridge. Each Gerber supported span was 36 m long and comprised six precast pre-stressed beams, with a variable depth equal to 2.20 m at mid span, sitting on Gerber saddles protruding from the main deck.

1.2 Main Deck

The geometry of the deck section has already been briefly described above, with its reinforcement being now herein discussed. Morandi [7] had noted that some parts of the deck were *essentially lacking longitudinal reinforcement*, a statement that seems

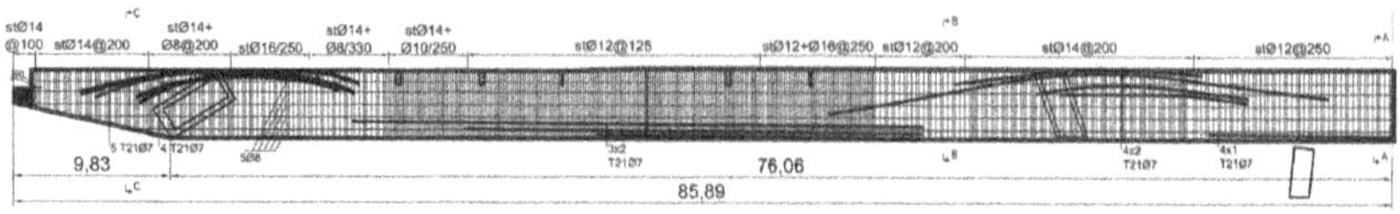

Fig. 3 Geometry and reinforcement of the main deck

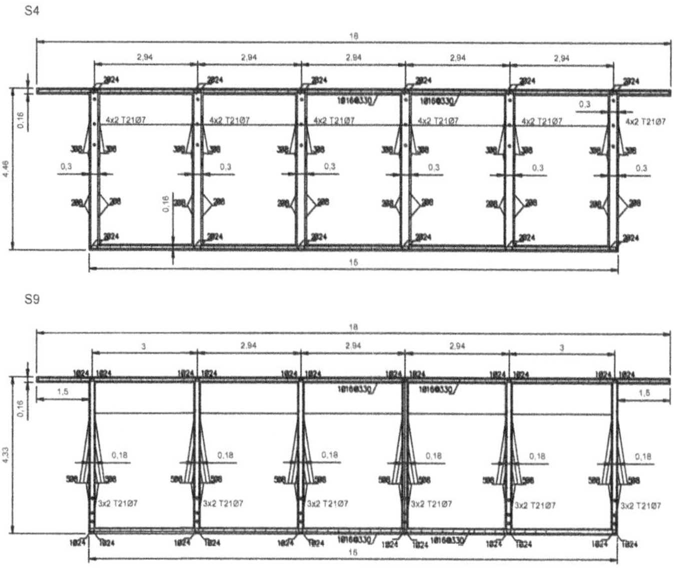

Fig. 4 Cross-sections S4 and S9 of the main deck

indeed aligned with the reinforcement quantities actually employed, especially if compared to today's standard practice.

From the original drawings, reproduced in Fig. 3 and Fig. 4, it appears that the continuous reinforcement provided in each beam was 4 Ø24 and 10 Ø8 mm bars; on a standard concrete beam section of 0.18 × 4.5 m, a reinforcement percentage of $\rho_s = 0.29\%$ is thus obtained. Further, it appears that 8 pre-compression cables with 21 Ø7 mm wires (total $A_{sp} = 646.5\ cm^2$) were located on top of the beam in a region of about 12 m on each side of the connection with the pier inclined strut and in the cantilever hanging out of the stays' connection.

In the central region, 6 similar cables (total $A_{sp} = 484.9\ cm^2$) were located at the bottom slab of the box deck section, in correspondence to each beam. The upper and lower pre-compression cables do not appear to be overlapping, but rather leaving limited portions of the deck reinforced by the ordinary reinforcement alone.

From a flexural point of view, simple hand-calculations indicate that the capacity of the deck section is adequate, without much conservativism, in the regions of maximum moments, both positive and negative. However, in the regions next to the point in which a zero-moment value is predicted (depending on the loading condition), the capacity is largely dependent on the positive effect of the compression force originated by the stays' force horizontal component, with limited ability of absorbing any moment inversion due to unexpected actions. It is evident that the deck would not have been able to resist even its own weight without the restraining action provided by the cable; for this reason, during the construction phases when the stays were not yet present or active, the presence of the temporary cables, later removed, was essential.

Shear reinforcement was provided in a rather customary fashion, with varying numbers of stirrups and different diameters. In regions close to the stay connections, it appears that Ø14 and Ø8 mm stirrups were provided at a spacing of 200 mm. In other parts, two Ø12 mm stirrups were provided at a spacing of 250 mm. Again, according to simple hand calculations, the capacity would seem to be in excess of the expected demand.

2 Previous Collapse Sequence Modelling

With a view to explore potential causes of the observed failure of the bridge, Calvi et al. [3] carried out an extensive modelling exercise, which led to the dismissal of some of the speculative collapse-triggering hypotheses being made in the immediate aftermath of the collapse (e.g. corrosion-induced elongation of the strands), demonstrating instead that a sudden failure of a bridge stay connections with the deck or the antenna (Fig. 5) constituted a much more plausible motive for the collapse of the bridge. Obviously, such stay failure may have been in turn induced by other accidental events (e.g. fall of a heavy steel coil from a truck that was crossing the bridge), possibly in combination with the presence of significant corrosion in some section of a stay.

As can be readily gathered in Fig. 6, the collapse sequence, as induced by the antenna-to-stay interface failure, starts by a torsional collapse of the deck in a section next to the pier strut and the subsequent falling to the ground of the corresponding supported span. For this reason, in this work, the flexural–torsional-shear capacity

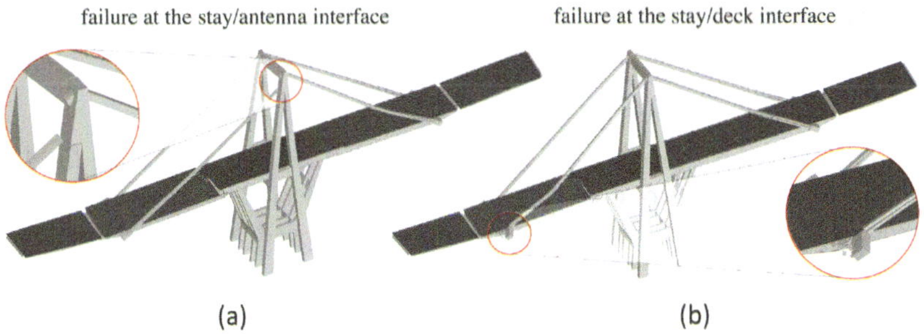

Fig. 5 Failure of a stay at the interface with (a) antenna and (b) deck [3]

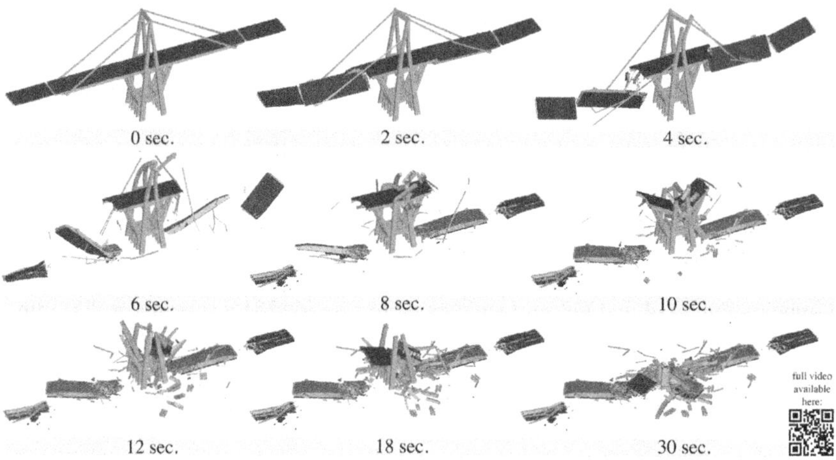

Fig. 6 Predicted collapse mechanism associated to a sudden failure of the connection between antenna and stay [3]

of the decks is further studied, through the employment of numerical modelling strategies (Finite Element Modelling - FEM) alternative to the approach considered by Calvi et al. [3], i.e. the Applied Element Method (AEM), with a view to confirm, or not, the results obtained by the latter.

3 Cross-section Analysis of Main Deck

In what follows, flexural, shear and torsional capacities of the main deck of the bridge are assessed through cross-section analysis, and then compared against the demands estimated by Calvi et al. [3] under three loading scenarios, (i) static vertical loads (i.e. after the completion of the different construction stages of the bridge), (ii) seismic loads (collapse limit prevention limit state), (iii) loss of a stay.

3.1 Flexure Capacity of Main Deck

The sectional analysis program Response-2000 [1], which calculates the full load-deformation response of reinforced cross-sections subjected to shear, moment, and axial load, was employed. For simplification purposes, and taking advantage of the double symmetry of the deck's cross-section, only one single I-beam portion of the complete section (which, as shown in Fig. 3 above is formed by a five-cell box girder) was analysed; the total bending moment capacity of the deck cross-section was therefore considered to be six times the capacity of the analysed single I-beam.

The capacity was evaluated considering the simultaneous application of axial and transverse loads, whilst torsion and in-plane bending/shear could not be considered. Three critical locations (Fig. 7), were considered for the verification of the bending moment capacity of the main deck cross-section; 1) at the support provided by the pier; 2) at the connection with the cable-stay; and 3) approximately between the pier-deck connection and cable-stay-deck connection (Sects. 1 and 2). Using moment–curvature analysis, both positive and negative bending moment capacities were estimated for the deck cross-section at the aforementioned locations, to consider potential moment demand reversal. The moment–curvature responses of the three identified critical sections are presented in Fig. 7.

The comparison between the bending moment demand and capacity of the main deck is illustrated in Fig. 8, from which it can be seen how the capacity foreseen by the original design is enough to cover the demand stemming from all the loading stages, as well as the increase originated by the consideration of the seismic loading. Even in the case of a stay removal on one side of the balanced system, it is clear that no longitudinal flexural issues would arise, when considering the equivalent beam to represent the entire deck section (i.e. neglecting torsion and bending/shear in the horizontal plane).

3.2 *Shear Capacity of Main Deck and Supported Spans*

The shear capacity estimates are based on a simplified version of the Modified Compression Field Theory (MCFT) [2], formulated by Vecchio and Collins [12], and discussed further in Sect. 4.1, below. Similarly to the case of the calculation of the longitudinal flexural capacity, above, the shear resistance of the main deck

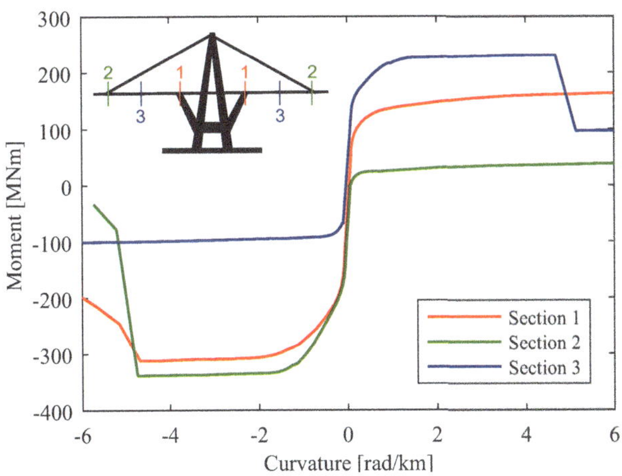

Fig. 7. Moment–curvature response of the considered deck cross-sections [3]

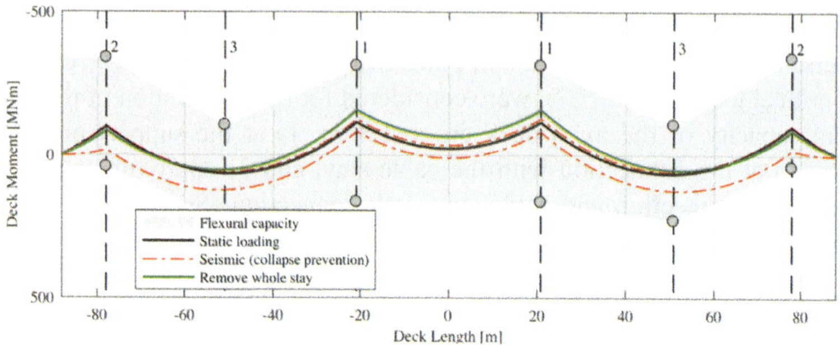

Fig. 8. Comparison of the bending moment demand along the deck with respect to the section capacities computed from moment curvature analysis [3]

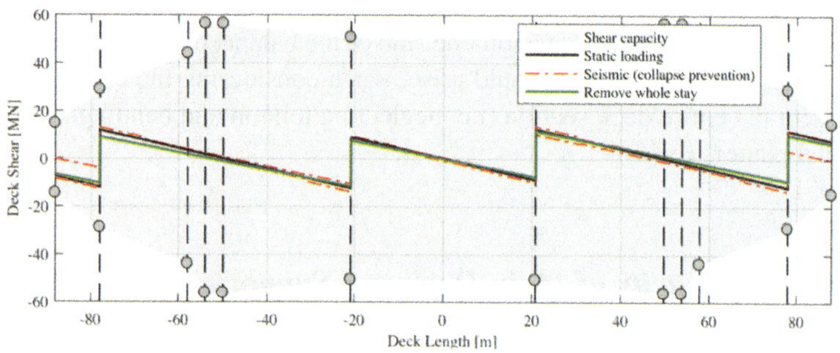

Fig. 9 Comparison of the shear demand along the deck with respect to the section capacities computed using modified compression field theory [3]

was computed at various points, still neglecting torsion and in-plane flexural/shear deformations, and is compared with the demand for the various loading scenarios,

As in the case of flexure, the shear capacity of the main deck is well above the maximum anticipated demands from both the static and the seismic loadings, indicating that it had sufficient reserve capacity. Furthermore, for the case of removal of the stay on the left-hand side, it can be seen that the shear demand in the deck over the left pier does not exceed the capacity, indicating that should a stay be removed from the system, no problems should be expected as a result of the increased shear demand (Fig. 9).

3.3 *Torsional Capacity of Main Deck*

The "Variable-Angle Truss Model" proposed by Rabbat and Collins [9] and implemented in the Canadian Concrete Structures Design Code [5] was herein employed. In this model, the cross-section is idealised using four parallel longitudinal chords,

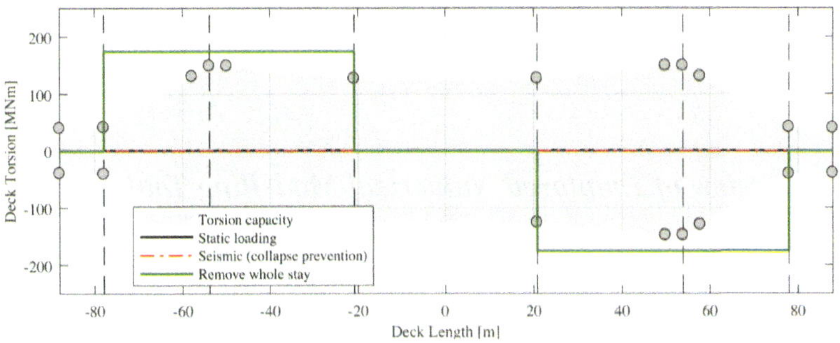

Fig. 10. Torsional capacity, computed both under normal and post-stay removal loading conditions, at various points of the main deck [3]

made of longitudinal pre-stressing steel, reinforcing bars and concrete. The chords are connected by four "walls", consisting of diagonally cracked concrete and transverse reinforcement. Moment and axial forces acting on the cross-section are resisted by axial stresses that arise in the chords, while shears and torsions acting on the cross section are resisted by shear flows that develop in the walls. In performing the calculations, a simplifying assumption was made,the shear flow generated by an applied torque was assumed to distribute only along the perimeter of the cross section, thus weighting only on the flanges and on the two most outer webs.

As before, the torsional capacity was computed for a number of sections along the main deck, and, as shown in Fig. 10, under the scenario of a stay removal, the torsional demand does exceed the torsional capacity. Whilst this is in-line with the progressive collapse sequence shown in Fig. 6, it is still a result that was obtained by making a few simplifying assumptions, which thus warrants the more in-depth modelling and analysis described in the subsequent Section.

4 Detailed Modelling and Analysis of Main Deck

The analyses discussed so far precluded the possibility of undertaking a detailed modelling of the main deck of the bridge. Indeed, on the one hand, the progressive collapse numerical study by Calvi et al. [3], summarised in Sect. 2, required the modelling of the entire structure, which effectively implies that fine discretisation and detailing could not be adopted, otherwise the computational burden would become unfeasible. The cross-section analyses described in Sect. 3, on the other hand, are by definition of a simplified nature, hence, again, consideration of reinforcement details and of simultaneous torsion, flexure, shear and axial actions is not possible.

In the current Section, an attempt was thus made to carry out a detailed refined modelling of the bridge deck, using the VecTor3 numerical modelling tool, with the objective of obtaining a definitive assessment on if it could or not resist the release of one of the stays (it is recalled that Calvi et al. [3], through the employment of a different structural analysis software, concluded that the deck could not cope with the release of one of the stays).

4.1 Overview of Employed Numerical Modelling Tool

VecTor3 [6, 11] is a MCFT-based program for nonlinear finite element analysis of three-dimensional reinforced concrete structures. The MCFT, briefly introduced in Sect. 3.2 above, represents a generalised approach for modelling the behaviour of reinforced concrete elements subjected to multi-axial loading conditions. It consists of a smeared, rotating crack model that treats stresses and strains in a localized average sense, and allows their reorientation as a result of changing load and/or material response. A key feature of the model is the examination of local behaviour at crack locations, therefore combining the average macroscopic representation of the material behaviour with the local shear stress-shear slip response of cracks.

The governing constitutive models of the MCFT were developed empirically from a series of reinforced concrete panel elements tested at the University of Toronto using the Panel Element Tester [10]. In particular, the results of the panel tests led to the development of compression softening and tension stiffening relationships that were incorporated into the constitutive compressive response for cracked reinforced concrete. Since its formulation, the MCFT has been corroborated by more than 200 tests of reinforced concrete panel elements, and a large number of tests for reinforced concrete beams and other full-scale elements.

The post-processor Janus [4], supporting the VecTor suite of nonlinear finite element analysis programs, was used for viewing and synthesizing analysis results as well as verifying model specifications.

4.2 Model Description

As discussed before (Sect. 1.1), the main deck is supported on four locations; two provided by the pier inclined struts at a distance of 42 m, and two provided by the cable stays at a distance of 152 m – clearly the latter cannot be considered as fixed, given that the deformation of the stays cannot be neglected. Hence, the static scheme of the deck could be considered to be that of a continuous beam on four supports with different stiffness (Fig. 11). In addition, taking advantage of the symmetric configuration of the bridge, it is possible to simplify the static scheme of the deck even further, as shown also in Fig. 11.

It is assumed that the stays behave elastically and the effect of the self-weight is neglected. Construction joints between adjacent segments were not modelled. In Fig. 12, the solid elements model developed in VecTor3 are shown. Close to 20,000 eight-node regular hexahedral elements were used to represent the concrete and 2,400 trusses bar elements were employed to model the longitudinal tendons (Fig. 13). Note

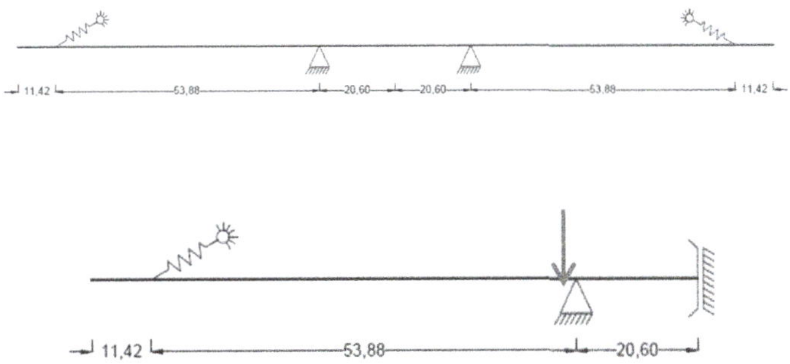

Fig. 11 Simplified planar static scheme of the bridge's main deck

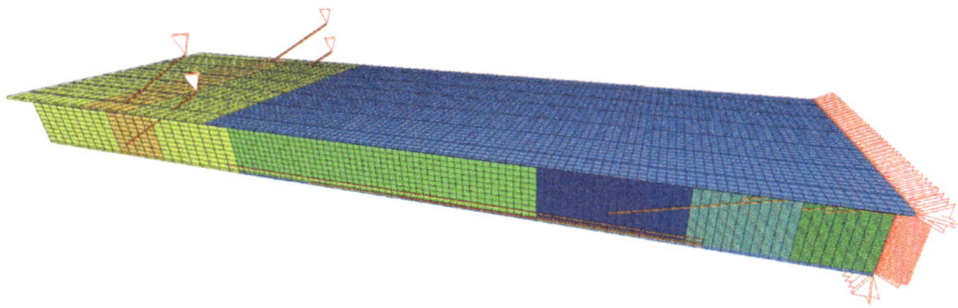

Fig. 12 VecTor3 solid elements model (different colours indicate different material models or properties)

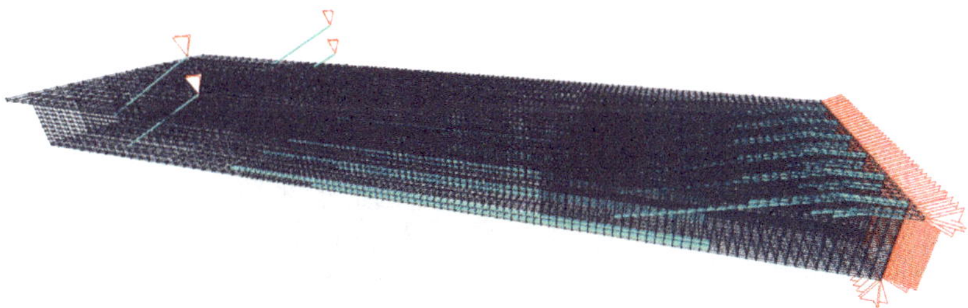

Fig. 13 Modelling of post-tension tendons

that the deck is assumed to be fixed on top of the support provided by the inclined strut.

A finer meshing was adopted in regions where higher concentration of inelasticity is expected following the release of one of the stays (i.e. closer to the stiffer supports), with sensitivity analyses being carried out to make sure that the adopted coarser meshing in less deformable parts of the deck was not impacting on the accuracy of the results. In regions of mesh transition, rigid constraints were employed to link the nodes of the smaller elements to those of the larger ones.

Longitudinal and transverse reinforcement were smeared within the concrete elements; different colours in Fig. 12 indicate parts of the deck with different reinforcement percentages. The part of the deck coloured orange is considered infinitely stiff, given the presence of the prestressed transverse girder (see Sect. 1.1), whilst the elements coloured yellow, where the smaller deformations are expected to occur subsequent to the release of a cable stay, feature elastic constitutive material models, in order to reduce computation time.

Loads applied to the deck model include gravity loads such as self-weight, permanent road loads and the jersey barriers (values of such loading are given in Calvi et al. [3]), whilst two point loads were applied at the top left edges of each of the six deck webs to represent the weight coming from the simply-supported Gerber beam spans (see Sect. 1.1).

4.3 Analysis and Results

The analysis consisted of gradually decreasing the support forces introduced by one of the stays (which was actually achieved by applying monotonically increasing loads in the direction of the stay, i.e. a sort of pushdown), thus introducing torsion in the deck, and allowing the monitoring of the ensuing deformations and the development of torsional resistance.

In Fig. 14, the deformed shape at the step of first rupture of the tendons is shown. As expected, and as outlined in Fig. 15, such first tendon rupture occurs in the more

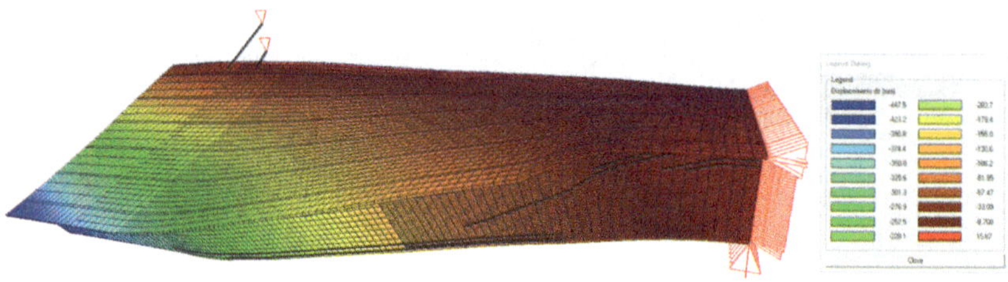

Fig. 14 Deformed shape at the step of first rupture in the longitudinal tendons

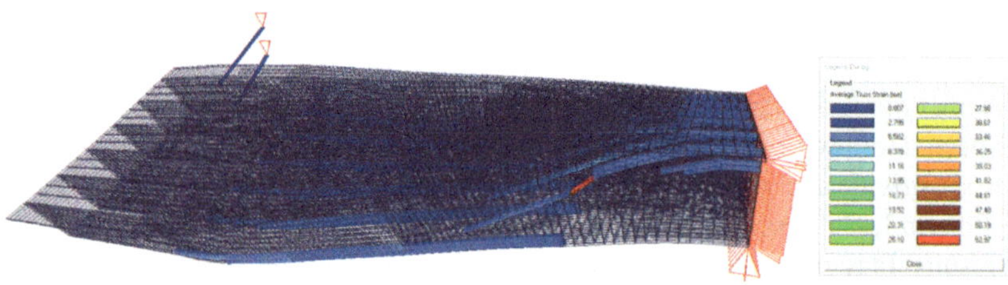

Fig. 15 Average strain in the longitudinal tendons

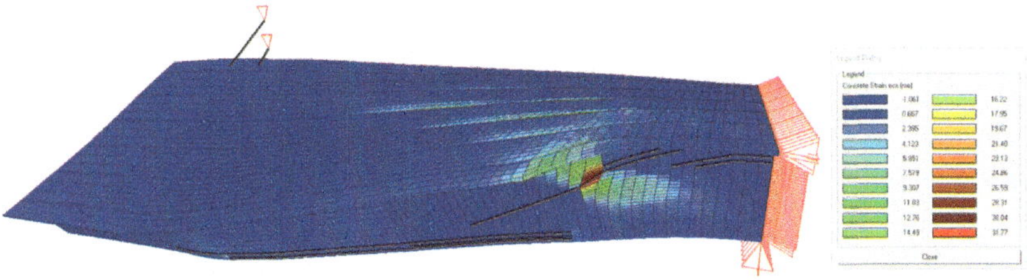

Fig. 16 Strain in the concrete along the longitudinal direction

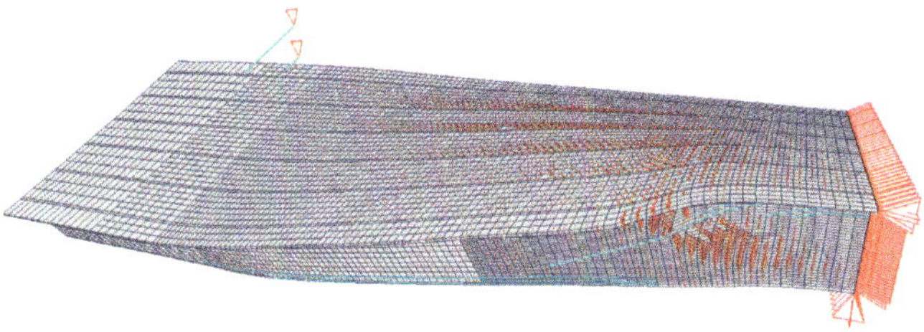

Fig. 17 Crack pattern

external web, where deformations are larger (see also Figs. 16 and 17, which depict concrete strain in the longitudinal direction and global crack pattern, respectively). In other words, the section where there is a higher concentration of damage is located on the right upper part of the deck, with the critical section being the end of the fourth layer of tendons (each layer is composed of two parallel tendons).

In Fig. 18, a plot of the deck's differential vertical displacement (measured at the stays-deck connection points) against the torsional moment generated at the deck is shown, where it can be observed that the maximum torsional resistance of the deck, based on the equilibrium of the system, is around 125 MNm (reached at a differential vertical displacement value of 250 mm).

Such torsional moment capacity compares well with the values shown in Fig. 10, and, is lower than the 175 MNm torsional demand estimated by Calvi et al. [3] in the event of a stay removal, thus implying indeed a collapse of the deck.

Always through the undertaking of static equilibrium calculations, it is possible to estimate maximum 5 MN shear force and 50 MNm negative bending moment values, which are well below the cross-section capacities presented in Sect. 3. This again corroborates the results obtained by Calvi et al. [3] regarding the torsional failure nature of the bridge collapse.

Further, the maximum value of differential displacement between the stays obtained here as being needed to induce the torsional collapse of the deck is well aligned with the 240 mm estimated by Calvi et al. [3], using their AEM modelling of the whole bridge, for the generation of a torsional collapse of the bridge deck.

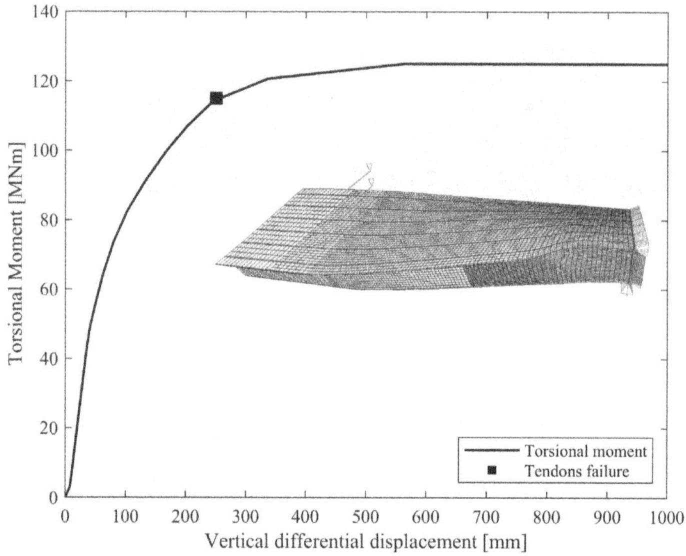

Fig. 18 Capacity of the deck with an increasing monotonic torsional load

Scrutinising further the above-described results that were obtained through the current modelling endeavour, one may note that the computed deck's torsional resistance is reached for a stay axial force of around 35 MN. Such value is less than the 39 MN stays' axial force value that the analyses carried out by Calvi et al. [3] showed to be needed in order for static equilibrium to be reached under the bridge's vertical loading. Once again this observation is congruent with the hypothesis of a torsional failure of the deck.

Put differently, in order to avoid a torsional failure of the deck, there would be a need for each stay to always be able to carry at least 39–35 = 4 MN, plus its self-weight of 1.25 MN, for a total minimum axial force value of approximately 5.25 MN. Hence, if one of the stays cannot carry any load, e.g. due to a rupture of its connection to the deck or the antenna, then a torsional collapse of the deck becomes unavoidable.

5 Conclusions

In this study, a more detailed analysis of the Morandi bridge deck was carried out with a view to verify if previous results and conclusions by Calvi et al. [3] regarding the collapse sequence of the bridge could be further corroborated. Such collapse sequence involved the torsional failure of the deck subsequent to the loss of a stay, for which reason a well-known Finite Element Analysis software tool that has been widely validated for the estimation of flexural–torsional-shear capacity of reinforced concrete structural members, was thus employed to further assess the torsional capacity of the deck.

The obtained results showed that in order to avoid a torsional failure of the deck, there is a need for the remaining stay on the same side of the collapsed one to be available to carry a minimum axial force value of approximately 5.25 MN, which effectively confirms the Applied Element Modelling results by Calvi et al. [3] indicating that the deck would not be capable of withstanding the loss of the support provided by one of the cable stays.

Further, the deck's differential vertical displacement capacity of 250 mm (i.e. maximum vertical deformation along its transverse direction that the deck can experience without collapsing) also seems to corroborate the considerations of Calvi et al. [3] on the fact that a slow progressive elongation of the stay, induced by progressive corrosion throughout the years, could not be considered as a main trigger of the collapse, since such large pre-collapse deformations would have been easily noticed.

Finally, and although not necessarily critical for what concerns the conclusions above, it is noted that variations (with the respect to the online available drawings used here to developed the numerical models) of e.g. tendons location and layouts, could generate concentration of damage in a section of the deck different from that shown in this work. We will thus continue to endeavour to acquire additional certainties regarding the actual construction details of the bridge.

Acknowledgements The authors would like to acknowledge the contribution of Gerard O'Reilly, Ricardo Monteiro, Daniele Malomo, Luis Alvarez, Maria Grazia Accurso-Tagano, Andrea Orgnoni and Federico Bellotti in different but related stages of this work.

References

1. Bentz EC (2017) *Response-2000*. Available at https://www.ecf.utoronto.ca/~bentz/r2k.htm (last accessed Nov 26, 2018)
2. Bentz EC, Vecchio FJ, Collins MP (2006) Simplified Modified Compression Field Theory for calculating shear strength of reinforced concrete elements. ACI Struct J 103(4):614–624
3. Calvi GM, Moratti M, O'Reilly GJ, Scattarreggia N, Monteiro R, Malomo D, Calvi PM, Pinho R (2019) Once upon a Time in Italy: The Tale of the Morandi Bridge. Struct Eng Int 29(2):198–217
4. Chak IN (2013) Janus: A Post-Processor for VecTor Analysis Software. MASc thesis, University of Toronto, Toronto, ON, Canada
5. CSA (2014) Design of Concrete Structures. A23.3–14, Canadian Standards Association, Mississauga, ON, Canada
6. ElMohandes F, Vecchio FJ (2014) VecTor3: A. User's Manual and B. Sample Coupled Thermal and Structural Analysis, Technical report, Department of Civil Engineering, University of Toronto, Canada
7. Morandi R (1967a) Il viadotto sul Polcevera per l'autostrada Genova-Savona. L'Industria Italiana del Cemento, XXXVII, pp 849–872
8. Morandi R (1967b) Il viadotto del Polcevera dell'Autostrada Genova – Savona. Roma
9. Rabbat B, Collins MP (1978) A Variable Angle Space Truss Model for structural concrete members subjected to complex loading. ACI SP-55:547–587

10. Vecchio FJ (1979) Shear Rig Design. *MEng Dissertation*, University of Toronto, Department of Civil Engineering, 246 pp
11. Vecchio FJ (2019) VecTor3. https://vectoranalysisgroup.com/vector3.html. Accessed 26 Feb 2019
12. Vecchio FJ, Collins MP (1986) The Modified Compression Field theory for reinforced concrete elements subjected to shear. ACI J 83(2):219–231

Importance of "Heuristics" in Suspension Bridge Engineering and 1915 Çanakkale Bridge

Ersin Arıoğlu

Abstract Suspension bridges can be regarded as masterpieces of the engineering profession. Although they are conceptually clear cut 5-piece load-bearing systems which are highly hyperstatic and their sizes are enormous. They undergo large displacements under loads, have nonlinear behavior and are sensitive to horizontal loads, especially wind loads. Suspension bridges are the most elegant, aesthetic and relatively economic structures of our civilization. The three decision methods to solve problems of engineers are "logic", "probability" and "intuition". Intuition, namely "heuristics" can be a vital influence in their decisions. Especially bridge designs are based on mathematical models, which take into account known patterns of physical behavior, but there are potentially a good many unknowns and uncertainties, which means heuristics is very useful and even required for short cut of the calculations and analysis. As the numbers of suspension bridges are increasing, a quite large database is becoming available for studying suspension bridges. At the same time, there has been a challenge to span longer distances, pushing the envelope of the engineering experience into new territory. This paper explores practical mathematical expressions obtained through regression analyses to predict easily key design parameters of long span suspension bridges such as main geometric dimensions, material quantities/qualities and dynamic properties for preliminary design calculations and estimations in order to satisfy. A large design parameter database matrix for 20 long span suspension bridges was collected to bring out heuristic approximations through regression analyses. Finally, these regression models are used to examine the design parameters of 1915 Çanakkale Bridge Project, which will break the longest span record with a main span length of 2023 m and the tallest tower record with 318 m (IP Point). It was observed that the dimensions, mass distributions and material qualities selected for the design of 1915 Çanakkale Bridge agree with the findings of this study.

This paper is reproduced based on the keynote speech named "Importance of 'Heuristics' in Suspension Bridge Engineering and the 1915 Çanakkale Bridge" which was performed in the IABSE - The Third Istanbul Bridge Conference (IBRIDGE) on 06/11/2018

E. Arıoğlu (✉)
Chairman of Yapı Merkezi, İstanbul, Turkey
e-mail: ersin.arioglu@ym.com.tr

P. Gülkan et al. (eds.), *Developments in International Bridge Engineering*, Springer Tracts on Transportation and Traffic 17, https://doi.org/10.1007/978-3-030-59169-4_2

Keywords Suspension bridge · Profound engineering · Sustainability · Uncertainty · Engineering logic · Art in structural engineering · Heuristics in suspension bridge · 1915 Çanakkale bridge

1 Introduction

1.1 Objective and Scope

In this paper, one of the masterpieces of engineers, namely *suspension bridges*, are examined to generate new long-span suspension bridge heuristics that may serve for *validation* of some key design parameters of these structures such as main geometric dimensions, material quantities and dynamic properties. These heuristics are planned to be utilized to validate the design parameters of the 1915 Çanakkale Bridge which will be the longest span suspension bridge with the tallest towers in the world when completed. To reach this objective, firstly, some ideas related to "*profound engineering*" which is needed to develop *sustainable* solutions on the existential problems of our civilization are expressed. Secondly, the *mentality* of engineers, their decision making and analysis methods, the "*importance of intuitive thinking with heuristics*", and the definitions of the decisional processes that transform engineering into art are presented. Then, general characteristics of the long-span suspension bridge behavior and features of the 1915 Çanakkale Bridge are discussed. Next, a benchmark data set covering a matrix of "20 × 120" parameters is gathered from the longest suspension bridges to identify relationships between various bridge parameters in an effort to generate some heuristics related with suspension bridge design.

Çanakkale Project is being executed as a fast-track project due to the requirement of the Client aiming to put the Project into service before 2023 and because the project is being executed under a "Public Private Partnership" *(PPP)* scheme. For this reason, *"Testing Methodology for Design Decisions through Heuristics"* was specifically developed by Yapı Merkezi, for quick benchmarking, validating the results of the "*design work*" and as well as to satisfy the Investors' reservations for the *"1915 Çanakkale Bridge"*. After the brief explanation of aesthetics of suspension bridges generally, 1915 Çanakkale Bridge is discussed specifically. Lastly, the concluding remarks of the study are presented.

1.2 Profound Engineering and Sustainability

Today, humanity is trying to experience two different civilizations at the same time. The first is the "*physical*" that we have been building for tens of thousands of years. The second one is the "*virtual*" that we set up in our computers, on the internet. It cannot be claimed that these two civilizations are in harmony with each other. Therefore, there are some problems that cannot be overcome yet. The most important of these are; "*inequality*", "*abuse of power*", "*environmental pollution*" and as a result "*climate change*".

By 2030, it is estimated that over the six billion people will have access to the internet and that we will be able to carry out operations and calculations approximately "*50 times*" faster than today [1]. This improvement has the potential to solve many of the aforementioned problems by making it easier for us to share information and "*cooperate*" with one another to achieve shared goals. On the other hand, it may also be the case that we create more "*chaos*" and "*entropy*". In short, "*our planet's future is completely dependent on humankind's intentions, the wisdom in their hearts and the way they make use of the potential they have...*".

The concept of "*competition*" today has been surrounded by the ideas of *learning speed* and *solution partnership.* Our civilizations expect us to make products that require less resources, but can last longer, that are stronger, faster, lighter, and gain absolutely more "*green*" and more "*aesthetic*" insights. In short, the world demands "*profound engineering systems*" from us.

To overcome their problems, people have learned to organize as *governments* and *organizations.* In the last fifty years, it is seen that developing more *"project-based"*, case-by-case solutions in *"compact"* teams to solve these problems produces more *"efficient"* results. The bright stars of today's generation will be those that are able to work effectively in *"project-based"* organizations.

The world is preparing to build a new civilization *uniting* both its physical and virtual aspects. This new civilization will prioritize achieving *"sustainable"* solutions. Humanity's *"strategy"* on this matter will probably be to *"increase the intelligence"* of each object by turning it into an *"internet item"* and to reinforce sustainability with the *"big-data"* they obtain.

In a broader sense, the concept of *"sustainability"* includes *"terms"* such as permanence, protecting the rights of future generations, endurance, resilience, and not giving up in the event of failure. Its main source is *nature.* It targets living *in harmony* with nature, *"to take as much as we give"*. In short, it recommends humanity to make use of nature as a *mentor* and not as a *"resource"*.

As learnt from the great mentor, Mother Nature, *ecosystems* that have been on our planet for at least three billion years have implemented many different and complicated strategies in order to survive and thrive in their environments. Even though an ecosystem does not necessarily leave behind a culture or written documents, it still finds ways to ensure its survival. In the last 30 years, mathematical models of complex systems, fractals, and chaos theories were developed in an attempt to understand these phenomena. The reward for sustainability is *"life"*. And the tool box for this will be *"profound engineering"*.

1.3 Mentality of Engineers

Sustainability is something that engineering, but only *profound engineering,* can achieve.

Engineers try to fulfill the needs of people and harmonize humanity with nature. There is no *"single solution"* for human needs. There can be a lot of solutions that are alternatives to each other. Engineer is the one who brings to light these alternatives and identifies the most *"effective"*, *"sustainable"* and *"beautiful"* option under the existing constraints. Engineers are skilled enough to define targets and to reach them in tolerable limits. Engineers make everything measurable; they measure and line everything up. They use limited resources efficiently. With these special talents, engineers are able to create economical wealth by continuously increasing the welfare of communities. In other words, engineers question, think hard and build *"civilizations"* and its *"systems"*.

They do not aim to make *"discoveries"* like scientists; but to design and build from *nothing;* in short, they aim to *"invent"*. Engineers are aware and apply science very well for their work. They can exist with their experiences, *"intuitions"*, aesthetic sense, and with limited empirical knowledge. As a matter of fact, many prestigious bridges were built in the Middle Ages and earlier, long before Newton's law of gravitation.

Engineering marvels can be seen in everywhere, although they are not often recognized as such by people. However, when a building collapses or a bridge is damaged, only then does society and media discuss the works of engineers, coming to realize their existence.

Engineering works are not created by just one engineer, rather they are products of *"huge collaborations"*. Every engineering work, even the newly built ones, contains knowledge and experience of thousands of years. Each one is built by taking into account the previous *"similars"*. Who can say that the Golden Gate Bridge was not inspired by the Brooklyn Bridge and those who built the Akashi Kaikyo Bridge did not learn from the Tacoma-Narrows Bridge? Like all knowledge passed on cleverly from previous trials to the new ones, theories and formulas flow from one bridge to another. The equations, formulas, and ideas of Pythagoras, Al-Jaber, Newton, Euler, Laplace, Lorenz, Napier, Steinman, Brown and many other scientists and engineers are blended within the calculations and heuristics of the designers and builders of each suspension bridge. That's the reason why engineers are *humble* and most of the time 'invisible'.

Engineers use *3 sets of criteria* that interact with and are inseparable from each other.

- *"Laws of nature":* This set is oriented on using the minimum amount of material and yet providing sufficient resistance to all effects of *"nature"* and *"function"*. It measures the *"performance"* of the work and questions its *"safety"*. Such criteria are universal, objective, and *"universally applicable"*.
- *"Economic conditions":* Economic conditions arise from *"geographic* and *social"* factors. These are the subjective criteria that make the cost benefit analysis of the work, measure its tolerable *"economic"* boundaries, and consider the *political will.*
- *"Aesthetic criteria":* These can be grouped into two. The first one is the criteria based on the intent of those who design and construct the artifact, the second

one is the expectations of those who use the artifact. These criteria question the *"elegance"* in the solution made under the constraints of nature and economic conditions. Mostly, they include the desire of the engineer to *"surpass"* previous projects by producing a better, more aesthetically appealing work. Aesthetic senses can be universal, but can also be influenced by the local cultures, especially *"history"*.

In short, engineers seek for systems with the *"minimal completeness; maximal applicability"* principle [2], which is the mathematical expression for aesthetics. In other words, they are after systems with *"just enough redundancy; maximum benefit"*.

Historical artifacts that are seen today are examples of how *"productive"* ideas were applied *"in a good way"*. For this reason, they continue to be regarded as *"aesthetic"* today, as was the case when they were built. The streams of aesthetics that have reached today display the importance of *"heuristic"* thinking in engineering. It proves that it is possible to reach the *"better"* over time, with designs foreseeing the evolution of flows. Master engineers reach *"aesthetic integrity"* by removing all unnecessary entities in design and by increasing the *"benefits"* of each remaining element in it. Like Sinan, the Architect who was the chief Ottoman architect and civil engineer and was responsible for the construction of more than 300 major structures during sixteenth century (Fig. 1).

In general, engineers seek answers to *two main questions:* 1) Is the *"technical"* capacity what we have, sufficient for the solution? *(Technical feasibility)* 2) Does it fit our budget? *(Economic feasibility)* If the answer to these two questions is *"yes"*, then there is a solution in *"hand"*. At this point, a *third* question *must be* asked: *"Where is the better one?"* This question leads classical engineering to profound engineering and the answer is the genuine reason of the emergence of *"profound engineering"*.

The management model that is becoming more and more important in the world of civil engineering is *"Project based Management"*. The project is a *"work package"* with determined targets, budget, duration and performance criteria that will be performed *"once"*.

The journey of *"making"* projects is like sailing during stormy weather. In this adventure, the weather conditions represent the 'conjuncture', *"uncertain"*; waves symbolize 'human behaviors', *"problematic"*. All resources required for the project are limited. The only compass for such an adventure would be the *"questions"*. Sound engineering questions ensure *good* answers and eliminate lack of knowledge. *In short,* the power for setting sail is found in the answers to these questions.

Engineer can ask questions to himself, his colleagues, his structure, the nature surrounding the structure, the community, the owner, the end-users, there is no limit... But every question will consume more from the limited budget of resources. The limit is proportional to the insufficient resources allocated to the project. Besides, if the road map of the *"problems"* that one wanted to be solved is full of *"uncertainties"* and *resources* are *"insufficient"*, then this is a *"pure engineering"* problem and it

Fig. 1 Selimiye Mosque – Edirne *(1575)* by Architect Sinan

requires the mentality of engineers. In short, it is needed creative engineers, with other words *"inventors"*.

2 Heuristics in Engineering

Engineers use three types of reasoning methods in order to solve problems. The 1st is *logic,* the 2nd is *probability;* which are descriptive, *numerical analysis* methods. The 3rd method used is *heuristics,* that is, *"intuitive thinking"*. These are *"little packages of intuitive ideas"* assumed to be based on *verbal*, common sense, which are formed by pieces of information such as experiences, guiding principles, and strategies (Fig. 2).

The imposing and *dual nature* of Aristotle's logic, *"either good or bad",* has dominated mankind's way of thinking for centuries. It has lost its power at the beginning of the twentieth century, as *mathematical logic* improved [3].

Probability mathematics came into the picture in the seventeenth century, in parallel to mathematical logic. Probability theories managed to emerge in science and everyday life, using statistical and experimental methods. As the science of psychology evolved, it became obvious that probability mathematics and the human mind were like the two sides of a medallion, complementing each other.

In the Western world, the centuries old term, *"heuristics"*, was derived from *"Eureka"*, Archimedes' famous saying. The term *"heuristics"* was recognized in

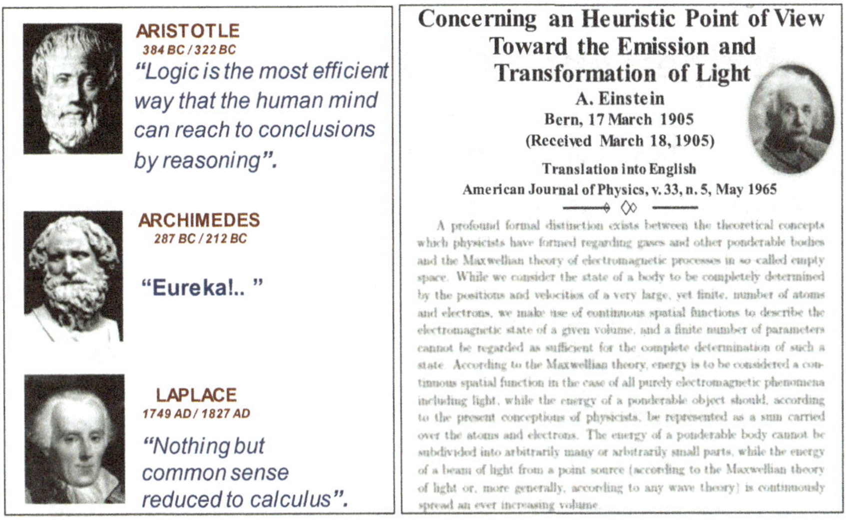

Fig. 2 Reasoning methods and their pioneers

the field of *science* in the beginning of twentieth century. Einstein used the term "*heuristic*" in the title of his paper, which was published in 1905, and lead to the way of Nobel Prize to him in 1921. After several years, Einstein was saying that; the theory presented in 1905 was *"incomplete but nevertheless very useful"* [4].

In 1970s, *Tversky* and *Kahnemann* explained that "Human kind think via their two systems". *System I*, the *"heuristic"* way of thinking, and *System II*, the *"numerical"* way of thinking. With these studies, Kahnemann earned the Nobel Prize in Economics at 2002 [5]. To summarize the systems of Kahnemann briefly:

System I: Heuristic, fast as reflexes, needs too little brain effort. On the other hand, it may evolve into a fixed opinion, is specific to the person, time and environment and cannot be used everywhere unless adapted. While it provides *"good enough"* solutions under the existing conditions, it is unable to provide the *"best"* solutions. Consequently, it seems as if this method is a *"magical"* treasure of ideas, empowering the ability of engineering to provide solutions with limited resources. Also, computer researchers are focused to deeply understand the heuristics to improve *"artificial intelligence"* and *"expert systems"* in the last decades.

System II: Numerical, provides *"optimal"* highly accurate and precise solutions, no cause and effect gap in the processes, errors can be easily eliminated by going backwards… On the other hand, this method requires focusing, as well as tiring and complicated calculations; consumes time, mental effort and financial resources.

Engineers make thousands of decisions during the design and construction phases. Every decision contains two different stages: *"judgement"* and "choice" (Fig. 3). Most of the time, these processes are full of *uncertainties.* Mainly five typical sources of *uncertainties* are: *Data insufficiency, uncertainty of the future, mismatch of numerical models with reality, variability of key system parameters and human errors.*

Judgement + Choice ➡ Decision[5]

If the event, with all the alternatives, reasons and solutions, together with the possibilities;[6]

- **is known,**
  - The case includes only "***RISK***"s.
  - **Statistical** thinking is ***'sufficient'*** for an ***'optimal'*** / best solution.
- **is not known,**
  - The case includes "***UNCERTAINTIES***".
  - **Intuitive** thinking *(heuristics)* is **necessary** and useful for the ***'best possible'*** solution.

Fig. 3 Judgement, Choice and Decision Process

Every source of uncertainty involves *"lack of knowledge"* and *"aleatory inherent in nature"*. System I and System II are implemented together while deciding in a sea of uncertainties. If everything is clear in a case, which is very rare, System II could be sufficient for an *optimal solution.*

To reduce the number of uncertainties, engineers constantly refresh their professional knowledge and *"improve their cognitive toolboxes"*. In their toolboxes, they keep specifications, quality control cards, testing methods, results of similar structures, ideas toward leading principles, and little packages of *"intuitive"* information. In other words, these are numerous *heuristics* that they collect via *System I* in their lives.

Engineers decide whether they acquire efficient solutions as a result of the calculations they make, or more as a result of their *"intuitions purified"* from their experiences. *Experience* is a learning curve; it is never *perfect* or *absolute.* Master engineers know the limits of their experiences very well. Using this experience via intuitions is the *craftsmanship* part of engineering. Conversely, using it via *science* and *aesthetics* is the *artistic* side of engineering.

Engineers hear the great symphony of the cosmos first through ***'science'***, feeling the mysteries of nature in their minds and muscles. Then, they listen to the melody, which is the result of the ***'stream'*** of events. They run the *'project'* according to this *'melody'*, organize it *with their intuitions,* form it *with their 'analyses'* and present it to us as a unique piece of *"music"*, that is frozen in space, like a ***'piece of art'***.

A related meaningful quote is *"Music is the hidden arithmetical exercise of a mind unconscious that it is calculating"* by *Wilhelm Leibniz* who is the first theoretician of computers.

3 Suspension Bridges and 1915 Çanakkale Bridge

3.1 General

There are around *15 million* bridges on nearly *150 million km* of roads, actively in use throughout the world. If these bridges are classified according to their main *load-bearing types,* it will be seen that there are *4 types* of bridges: *Beam, Arch, Cable-stayed,* and *Suspension* bridges (Fig. 4).

Another grouping based on the length of *bridges* is: *Short-span (less than 100* m*), mid-span (100 –600* m*), long-span (600–1200* m*),* and *very long-span* bridges *(longer than 1200* m*).* Today, the maximum spans that beam and arch bridges can reach up to are respectively 330 and 552 m, which is considered *mid-span.* Bridges that can be considered as *"long-span"* are only *cable* supported bridges. Bridges with a span of over 1200 m that are considered *"very long-span"* are merely *"suspension bridges"* today.

Apart from being magnificent and mysterious, *suspension bridges* are clear-cut and simple 5-piece load-bearing systems (Fig. 5). Suspension bridges are highly hyperstatic, their sizes are enormous; they make large displacements under loads, they have *nonlinear* behavior. They are sensitive to horizontal loads, *"especially wind loads"* that result from compressible air movements; also, they are the most elegant, aesthetic and relatively economic structures of our civilization.

3.2 Background of 1915 Çanakkale Bridge.

The span length of *suspension bridges,* which are the fastest developing type of bridge, has increased by nearly 10 times in the past 150 years. If the economic *and practical* limitations are left aside, there is no *"technical"* limitation to building a suspension bridge *10,000* m long, with the *"knowledge"*, *"materials"* and *"technologies"* available today [7].

Theoretical studies of the last 50 years and construction of suspension bridges with spans longer than 1500 m have proven that deck designs using *"multiple boxes"* showed sufficient resistance against wind loads within economic limitations. Following this solid discovery, the answer to the question of *"which project would beat the 2000* m *long span limit and when would that be"* was sought after. Indonesia, Japan and Norway were among candidates, but Turkey wasn't [8].

Fig. 4 Record bridges depending on bridge type

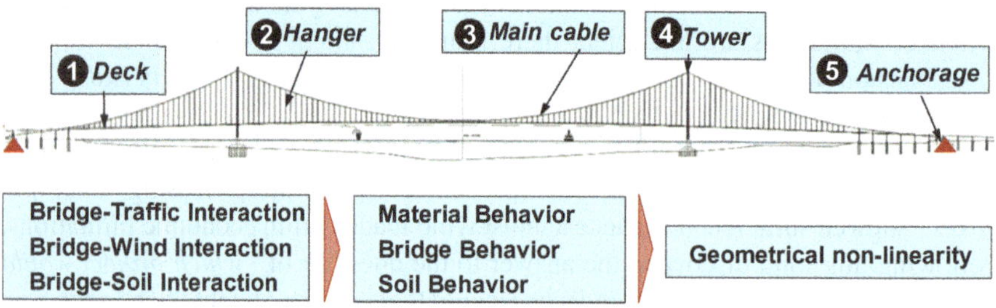

Fig. 5 Main components of a suspension bridge and critical considerations in design

However, General Directorate of Highways *(KGM)* in Turkey had been making research and collecting information about suspension bridges since the 1960s. They had two beautiful bridges, which were *good examples* of 2nd generation bridges, constructed in *1973* and in *1988* on the Bosphorus, in Istanbul. In *2016*, they completed and began operating two very significant, long-span bridges *(Osmangazi and Yavuz Sultan Selim with main span lengths of 1550 and 1408 m, respectively)* with the Build-Operate-Transfer *(BOT)* model, one after the other. It was an *"outstanding"* success to be able to put *two bridges* of this size into operation, within the same year. Soon after, KGM immediately went out to tender for a bridge that would be the longest-span suspension bridge in the world, the *"1915 Çanakkale Bridge"*, aiming to break ground on 18 March 2017 (Fig. 6).

Approaching the 2000 m span limit with the Akashi-Kaikyo Bridge [9] - Japan, the success of the multiple-box structure system was recognized with the Yi-Sun-Shin Bridge - South Korea, the robust findings gathered during the research on the Messina Bridge - Italy, and successful experiences gained with the Yavuz Sultan Selim Bridge - Turkey and Osmangazi Bridge – Turkey, encouraged KGM and the interested parties to make a record attempt for "1915 Çanakkale Bridge" (Fig. 7). Main features of Malkara–Çanakkale Motorway Project including 1915 Çanakkale Bridge, [10–12] presented as a group

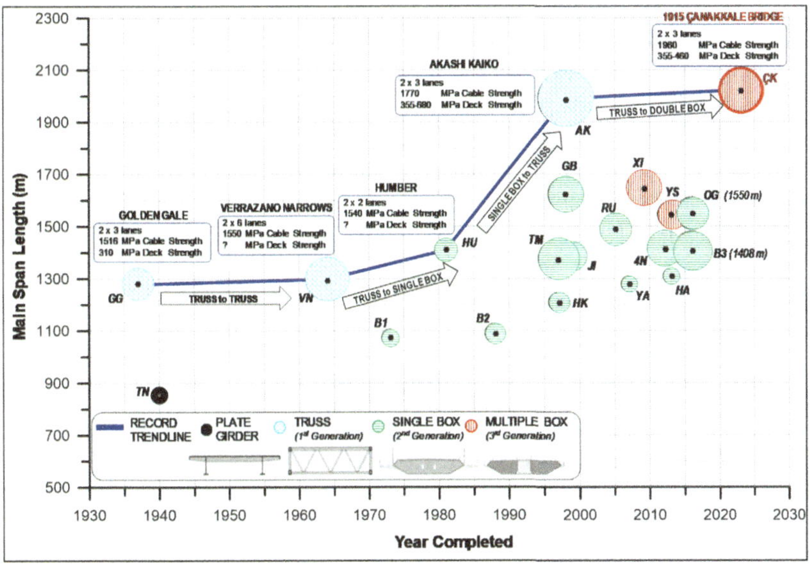

Fig. 6 Main span length development in suspension bridges. **Abbreviations: ÇK:** 1915 Çanakkale; **AK:** Akashi; **XI:** Xihoumen; **GB:** Great Belt; **OG:** Osmangazi; **YS:** Yi Sun Shin; **RU:** Runyang; **4N:** 4th Nanjing; **HU:** Humber; **B3:** 3rd Bosphorus; **JI:** Jiangyin; **TM:** Tsing Ma; **HA:** Hardanger; **VN:** Verrazano Narrows; **YA:** Yangluo; **GG:** Golden Gate; **HK:** Höga Kusten; **B2:** 2nd Bosphorus; **B1:** 1st Bosphorus; **TN:** Tacoma Narrows; **ME:** Messina (Bubble area relatively represents ***"Cable + Hanger + Deck"*** Weight)

Fig. 7 General view of 1915 Çanakkale Bridge

in Fig. 8. From then on, the international suspension bridge engineering world's *attention* has been focused on Çanakkale.

Symbolizing the successes of the Republic of Turkey completing its first 100 years, the 1915 Çanakkale Bridge is dedicated to *"peace"*, through remembering the martyrs who made Çanakkale impassable, in World War I. It is also a monument *dedicated to the progress of* global civilization. Together with the nature of its investment character and its design criteria, are extraordinary responsibilities for the investors/builders. This awareness has *"interlocked"* all the stakeholders, keeping them on *"targets"*.

4 Long Span Suspension Bridge Behavior

4.1 Performance under Static Loads

The duty of suspension bridge engineers is to ensure that the bridge, displays the required *performance* under all kinds of *static* and *dynamic* loads. As the spans of the bridge become longer, the rigidity of the deck decreases quickly and it becomes a "slender" element. The main cable that provides the main rigidity of the bridge progressively becomes a *"heavier backbone"* that bear most of the main vertical loads. More costly heavier cable will increase the cost of the other elements of the bridge as well. The feasibility and sustainability fall into danger. To overcome this, the engineers adapt some of the measures below [8]:

MALKARA - « ANAKKALE MOTORWAY FEATURES

OWNER	: MINISTRY OF TRANSPORT & INFRASTRUCTURE; GENERAL DIRECTORATE OF HIGHWAYS *(KGM)*
ENGINEER	: TEKFEN & T ENGINEERING
SPV INVESTOR	: ÇANAKKALE OTOYOL İŞLETME A.Ş. *(ÇOK A.Ş.)*
EPC CONTRACTOR	: DAELIM, LİMAK, SK E&C, YAPI MERKEZİ *(DLSY JV)*
DESIGNER	: COWI *(Bridge)*, Y‹ KSEL PROJE *(Motorway)*
IDV	: ARUP & AAS-JAKOBSEN

88 KM MOTORWAY
1 SUSPENSION BRIDGE (770m - 2023m - 770m = 3.563 m)
2 APPROACH VIADUCTS (1.045m)
2 O & M CENTER
4 MOTORWAY VIADUCTS (2.694m)
4 SERVICE AREAS
7 TOLL COLLECTION
11 UNDERPASSES
12 JUNCTIONS
30 PORTAL UNDERPASS
42 OVERPASSES / 216 CULVERTS

The Investment Character
(Requests of the Client)

Monumental,
Unique,
Ahead of its time,
Respectful to the history of Çanakkale Strait,
Honors 100th anniversary of the Republic of Turkey *(2023)*.

Main Design Criteria
(Requests of the Investor)

Functional & Safer,
Durable with High Quality,
Economic & Constructible,
Aesthetic;
Beauty / Distinctive architecture / History and environment conscious / Public acceptance / Integrated with the existing systems / Adaptable to the changes in time.

1915 « ANAKKALE BRIDGE FEATURES

Gravity Anchorages type
Caisson Foundations on Inclusion Piles

Sag Ratio	: 1 / 8,9
Concrete	: 500.000 t
Cable	: 35.000 t
Steel	: 88.000 t

Fig. 8 Malkara-Çanakkale Motorway Project features in detail

- The highest *"cable sag ratio"* is chosen, without affecting the rest of the bridge's qualifications.
- *Highest strength* cables are used for the production of the *main cable.*
- Safety factors and stress limits are defined *carefully.*
- The deck weight is reduced by the use of high-strength materials and appropriate *aerodynamic forms.*
- *Lighter* materials are preferred for the surfacing, fittings, and equipment of the deck.

Thus, balance between *"rigidity and economy"* can be achieved.

4.2 Performance Under Dynamic Loads

"Large enough" displacements are needed to *"dissipate energy"* under dynamic loads. On the other hand, only *"small enough"* displacements are allowed for structural integrity. Suspension bridge designs that can solve this *"dilemma"* are considered as successful designs. The result, as Kawada said: *"History of modern suspension bridge is solving the dilemma between economy and stifness"* [13].

In a structural system only with cables, its vertical and torsional frequencies are nearly at the same level of magnitude. In suspension bridges, the main cable is *dominant* in dynamic behavior like in static behavior, while the spans become longer. The vertical and torsional frequencies of the whole bridge almost resemble that of the main cables'. For this reason, the risk of the overlapping of vertical and torsional frequencies, and may cause undesirable aeroelastic instabilities in bridges [8], like *vortex-induced vibration, buffeting,* and *flutter*. Similar deficiencies are encountered in the history of cable supported bridges, during the construction or operation stages, stopping the operation or even sometimes causing the bridge to collapse.

The main precautions to increase the *"structural safety"* and to maintain *"aeroelastic stability"* in long-span suspension bridges are:

- Selecting of dimensions appropriately (*e.g. multiple deck with sufficient rigidity of cross girders)*;
- *Proper* distribution of the masses and rigidities and *testing* them afterwards;
- Increasing the *"energy dissipation"* and damping capacity;
- Introducing more *"redundancy"* to the system;
- Preventing incidents like *"resonances";*
- Equipping the bridge with active control systems *(dampers, buffers, guide vanes,* etc.*)*

4.3 Suspension Bridge Common Heuristics

Besides complicated numerical analysis, *"desired"* solutions have been reached by the use of many *"intuitive thoughts like heuristics"* during the design and construction of suspension bridges. *Heuristics* used by engineers can generally be grouped in five categories:

- ***In preliminary calculations:*** For instance; "*For steel with tentative yield point, the yield strength is equal to the stress value of 2‰ deformation*".
- ***For safety factors:*** For instance; *"In order to prevent cable stress to reach fatigue stress, a safety factor of 2,0 can be adopted".*
- ***Instructive:*** For instance*; "In third generation bridges,it is better to have more gap width than less gap width and small gap width than no opening at all".*
- ***For source allocations:*** For instance; *"Allocate sufficient resources to the weak(est) links.*
- ***Limiting the risks:*** For instance; *"The rabbit which has only one place to hide will be caught quickly..."* or *"You cannot make* "big" *changes to high technologies in one move..."*

It should be kept in mind that every *"intuitive decision"* is dependent on its purpose and content, time stamped and as "unique" as the applying engineer's "fingerprint". Anybody else can use them only if *"adapted"* with great care.

Some heuristics examples used for suspension bridges more frequently can be listed as follows:

- Main cable sag is chosen between ratios *1/8–1/12* [13–15].
- The ratio of the side span to main span should be between *0,35–0,50.*
- In third generation bridges, more gap is better than less gap; even small gap is better than gapless.
- If the form of the cable is considered as "catenary", the accuracy of the calculation results increase. When it is considered as "2nd degree parabola", the calculations shorten and become simpler. The difference between the results is negligible *(in terms of conceptual design)* [15].
- The radius of the tower saddle should be 8–10 times more than the cable's diameter [16].

5 Generated Heuristics for Suspension Bridges

5.1 Suspension Bridge Data Set

Utilizing *"similar"* experiences in bridge design is one of the most logical and efficient ways of *"producing intuitive ideas"*. As an efficient approach, a database was created to test the design parameters of the 1915 Çanakkale Bridge *(ÇK)*, using the data of *15 suspension bridges* with main spans longer than *1200* m. Additionally, two special bridges, Messina *(ME)* and Old Tacoma Narrows *(TN)*; and two Bosphorus bridges in Istanbul *(1st Bosphorus [B1] and 2nd Bosphorus [B2])* were included in this research to enrich the database. For each bridge *120* important design parameters was selected and a *database "matrix"* of *2400 cells (20 × 120)* was formed. With the best effort, only *87%* of the parameters were obtained by reviewing more than *1000* documents. Around over *100* regression analyzes were conducted to identify the relationship between parameters. Four regression analyses chosen from this research and their preliminary findings are summarized and presented in the article.

5.2 *Regression Results*

The ambition for crossing longer spans economically has caused the deck technologies of bridges to evolve: In chronological order; *1st generation*, truss; *2nd generation*, aerodynamic single-box; *3rd generation*, aerodynamic multiple-boxes.

5.2.1 Relationship Between Main Span Length and Total Main Cable Weight

In this study, when the change in cable weight is analyzed, according to the spans and generations, it is seen that *"Cable weight"*; decreases from the 1st generation to the 3rd, as technology develops to enable the construction of bridges with longer spans (Fig. 9). For instance, the cable weight of a *2000 m* 3rd generation bridge is *36%* less than the cable weight of a 1st generation bridge with the same span.

5.2.2 Relationship Between Main Span Length and Deck Depth

In another trial, linear regression models between the main span length and the *"deck height"* were set up for all three generations and it was observed that their slopes declined from the 1st generation to the 3rd (Fig. 10). In other words; it was seen that as *the spans became longer*, the deck heights of the new generations became more *efficient*. In *1st generation*, the deck height is *6%–7%* of the main span. In *2nd generation*, the deck height is *around 3%* of the main span. Lastly, it is seen that, the deck height is *less than 2%* of the main span in *3rd generation*.

5.2.3 Relationship Between Main Span Length and 1st Symmetrical Frequencies

Since the rigidity of the long-span suspension bridges reduces, they may be exposed to sudden and undesirable *"vibrations"* under dynamic loads such as wind, earthquake, and traffic. As Architectural historian writer Marcus Binney quoted "*Revolutionary*

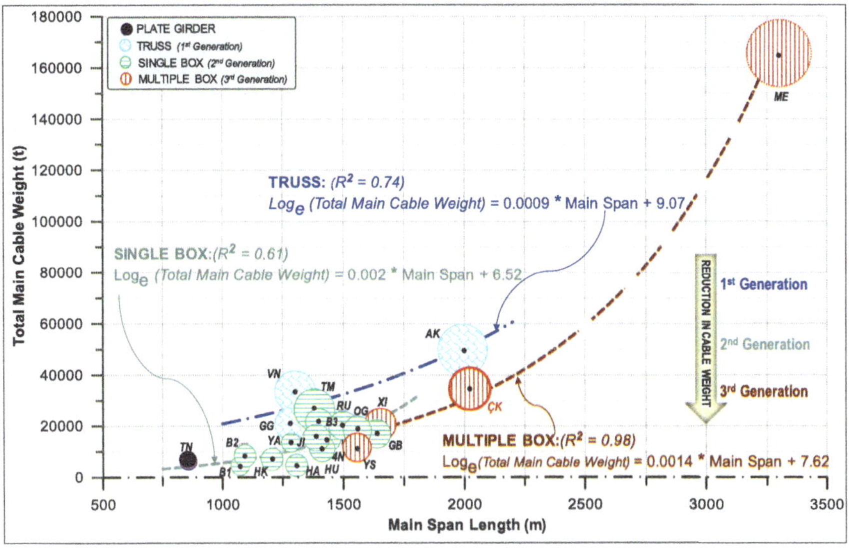

(Bubble area relatively represents ***"Cable + Hanger + Deck"*** *Weight)*

Fig. 9 Relationship between main span length and total main cable weight

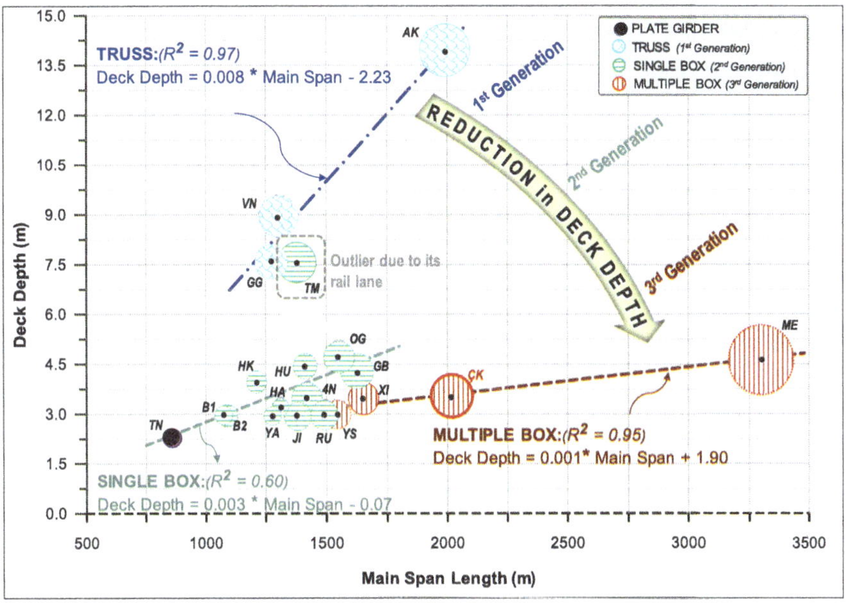

(Bubble area relatively represents ***"Cable + Hanger + Deck"*** *Weight)*

Fig. 10 Relationship between main span length and deck depth

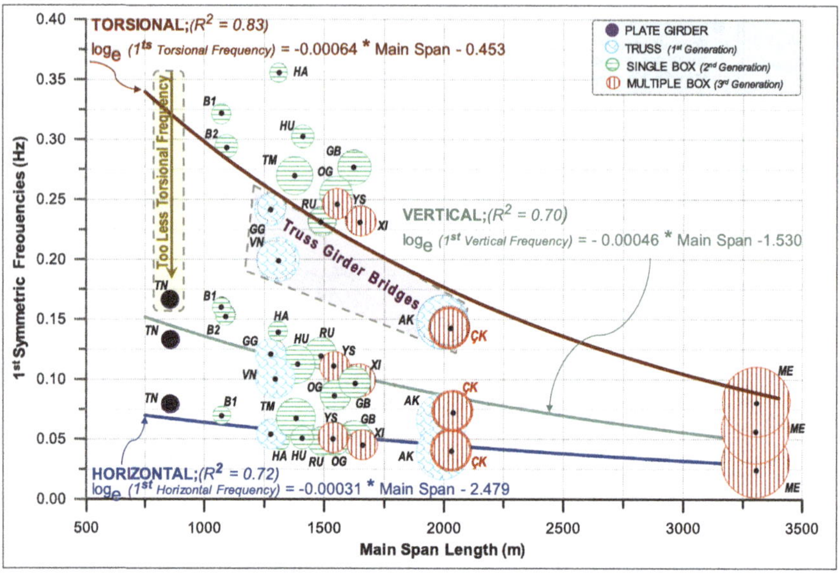

*(Bubble area relatively represents **"Cable + Hanger + Deck"** Weight)*

Fig. 11 Relationship between main span length and 1st symmetric frequencies

design (of suspension bridges) many times liable to give of bad (harmful) vibration. Then engineers make revolutionary inventions to avoid them".

It was confirmed that vibration frequencies *(the first symmetrical vertical, horizontal and torsional frequency values)* of the *Akashi (Truss)* Bridge, with a span of 1991 m, and the *1915 Çanakkale (Double Box Deck)* Bridge, with a span of 2023 m, are "similar" (Fig. 11). When the total weight of the deck, hanger, cable and tower materials are compared, it was understood that the 1915 Çanakkale Bridge could secure the same level of *"stiffness"* for aerodynamic stability with *"34% less material"*.

5.2.4 Relationship Between Main Span Length and Frequency Ratio

In suspension bridges, the *"ratio"* of the lowest symmetrical torsional and vertical frequencies affects the critical flutter speed. In the 1st and 2nd generation bridges, this ratio increases proportionally to the span and decreases in 3rd generation bridges (Fig. 12). In the 1915 Çanakkale Bridge, this ratio is kept around ~ *2.0,* providing the lowest recommended level in literature. This ratio decreases in 3rd generation bridges with spans longer than 2000 m that challenges to ensure the aeroelastic stability. In such cases, it will be necessary to design decks with wider gaps and/or use interactive wind control devices.

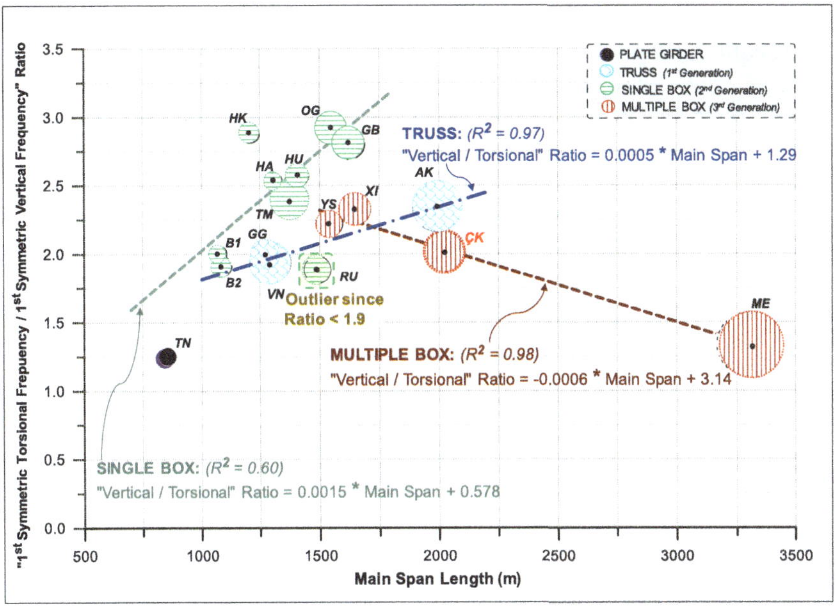

*(Bubble area relatively represents **"Cable + Hanger + Deck"** Weight)*

Fig. 12 Relationship between main span length and "1st symmetric torsional frequency / 1st symmetric vertical frequency" ratio

When the graphics presented here and the rest of the meaningful regression studies to be published in the close future are evaluated all together, it can be concluded that the dimensions, mass distributions and material qualities selected for the 1915 Çanakkale Bridge are *"accurately"* and *"properly"* designed. Therefore, it was seen that the *"Heuristic Method"*, used during the verifying and checking the design results, *"accelerated the process, enlightened the complex relationships between parameters and provided economy"*.

6 Aesthetics of Suspension Bridges

Fine art academician Robert Henri stated that; "*Good composition is like suspension bridge, each line adds strength and takes none away*". Any structure in particular suspension bridges, can only be as impressive and aesthetic as its design. Slenderness, harmony, proportions, symmetry, simplicity, colors, texture and interaction to the surrounding elements are the main components of *aesthetics.*

Bridges, especially monumental ones, cannot be designed independent from aesthetic concerns. But rather, the bridge must produce excitement through aesthetics in various levels and angles. If its aesthetic elements are of enduring value, the life of the '*edifice*' will be longer. This is because, the society will be more likely to protect such an edifice and give more attention to its preservation.

The enormous size of a suspension bridge does not diminish the significance of its visual message. In contrast, a long span suspension bridge must show that, it is able to bear all the loads in a naturally aesthetic form. The inherent, natural geometry of suspension bridges could be considered both aesthetic and successful because they serve the functionality in bearing the loads. By placing aesthetic elements in the design to convey, accentuate and enhance the "feeling" that the structure is resisting the loads gracefully and with ease considerably augments the natural beauty of these colossal structures. The designer of Golden Gate Bridge Joseph Strauss mentioned that; "*Ask of the steel, each strut and wire… What gave it force and power?*".

Aesthetic thought has many aspects; including safety, comfort, quality and service life of the structure, as well as its appeal to its users and its harmony with its environment. These aspects, together with profound engineering, should be masterfully blended into one continuous form to create a timeless masterpiece.

Accordingly, 1915 Çanakkale Bridge aesthetics had been carefully considered and many important features were taken into consideration such as; 2023 m main span for honoring 100th anniversary of Republic of Turkey, Turkish flag colors for towers and deck, tower's arch-shaped cross beams inspired from traditional Turkish architecture etc. (Fig. 13). It is believed that this bridge will be one of the symbols of contemporary civilization and become a "*stimulating*" artifact, with all its notable aesthetic and technical aspects.

Fig. 13 1915 Çanakkale Bridge – view of tower and cross beams. a. 1915 Çanakkale Bridge-Tower View 1. b. 1915 Çanakkale Bridge-Tower View 2

Fig. 13 (continued)

7 Concluding Remarks

After all, the concluding remarks can be enumerated as follows:

- The span length of *suspension bridges*, which is the *"fastest developing"* attribute of the bridge, has increased by nearly 10 times in the past 150 years. It is highly possible that over 3.000 m main span suspension bridges may be constructed in the next two decades.
- *"Heuristics"* are useful and powerful tools for engineers to find best possible solutions allowing them to make *productive* decisions in the presence of uncertainties. Each heuristic is as "*unique*" and like engineer's "*fingerprint*"; therefore use by others require great care.
- In the last 50 years, theoretical studies, wind tunnel tests and construction of suspension bridges with spans longer than 1500 m proved that *"multiple box decks"*, showed sufficient resistance against wind loads very economically by reducing deck heights to less than 2% of the main span.
- When the total weight of the deck, hanger, cable and tower materials are compared with Akashi Kaikyo Bridge which is current record holder for main span, it was understood that the 1915 Çanakkale Bridge could secure the same level of *"stiffness"* for *"aerodynamic"* stability with *"34% less material"*.
- Equipping the suspension bridge with *"active control systems" (dampers, buffers, wind guide vanes,* etc.*)* will become the norm in near future.

- Results of the regression studies revealed that the dimensions, mass distributions and material qualities selected for the 1915 Çanakkale Bridge are "*accurately*" and "*properly*" designed.
- *"The Testing Methodology for Design Decisions through Heuristics"* presented in this article was specifically developed for quick benchmarking and validating the results of the design works for the 1915 Çanakkale Bridge being built as a fast-track Project. The methodology is found to be *"useful", "effective"*. "The Testing Method" is expected to become widespread in the near future for conceptual designs of the long-span suspension bridges.
- 1915 Çanakkale Bridge will encourage future generations to make longer suspension bridges; thus bridging spans that were previously thought unbridgeable or considered uneconomical.

Acknowledgements The author would like to express grateful acknowledgement to; Prof.Dr.Ergin Arıoğlu *(R&D Coordinator)*, Başar Arıoğlu *(Chairman)*, Ömer Güzel *(General Coordinator)*, Dr.Burak Gökçe *(R&D Chief)*, Pınar Toru Şeker *(R&D Engineer)*, Yasin Karakuş *(Design Manager)*, Ümit Altınbaşak *(Chief Draftsman)* from Yapı Merkezi and Deniz Günsoy *(Expert)*, for their support in this publication. The opinions, findings, and conclusions expressed in this article are those of the author and do not necessarily reflect views of all organizations involved in the Project.

References

1. Schmidt E, Cohen J (2013) The New Digital Age – Reshaping the Future of People, Nations and Business, 315 pages. Alfred A. Knopf
2. King J (1992) The art of mathematics. Springer, 313 pages, pp 181
3. Whitehead AN, Russell B (1913) Principia Mathematica, 3, 1st edn. Cambridge University Press, Cambridge
4. Einstein A (1905) Concerning an heuristic point of view toward the emission and transformation of light. Annalen Phys **17**:132–148. (in German)
5. Kahneman D, Slovic P, Tversky A (April 1982) Judgement under uncertainity: heuristics and biases. Cambridge University Press, Cambridge April 1982
6. Gigerenzer G, Gaissmaier W (January 2011) Heuristic decision making. annual review of psychology, pp 451–482
7. Tang MC (2017) Super-long span bridges. Struct Infrastruct Eng 13(6):722–730
8. Brancaleoni F, Diana G, Faccioli G, Fiammenghi G, Firth I, Gimsing N, Jamiolkowski M, Sluszka P, Solari G, Volinsise G, Vulla E (2010) The messina strait bridge – a challenge and a dream. CRC Press, Taylor and Francis Group, 324 pages
9. Honshu-Shikoku Bridge Authority (1998) "Akashi-Kaikyo Bridge – Design and Construction of the World Longest Bridge, Second Edition"
10. Cam FS (2018) Kınalı-Tekirdağ-Çanakkale-Savaştepe Highway, Malkara-Çanakkale Section (Including 1915 Çanakkale Bridge, Build-Operate-Transfer, 1915 Çanakkale Bridge Presentation. Tekfen and T Engineering, IABSE I-Bridge November 5–6, 2018
11. Kroon IB (2018) "1915 Çanakkale Bridge – Meeting the Challenge", COWI, IABSE I-Bridge November 5–6, 2018
12. Küçükbekir S (2018) Kınalı-Tekirdağ-Çanakkale-Savaştepe Highway, Malkara-Çanakkale Section (Including 1915 Çanakkale Bridge, Build-Operate-Transfer Project Presentation. KGM, IABSE I-Bridge November 5–6, 2018

13. Kawada T (2010) History of the modern suspension bridge. ASCE Press, 246 pages
14. Chen WF, Duan L (2014) Bridge Engineering Handbook, 2nd Edition – Superstructure Design. CRC Press, 716 p.
15. Gimsing N.J, Georgakis CT (2012) Cable supported bridges, concept and design, 3rd edn. Wiley, 590 pages
16. Parke G, Hewson N (2008) ICE manual of bridge engineering second edition. Thomas Telford Ltd., London, 728 pages (2008)
17. Billington D (1996) Innovators: the engineering pioneers who transformed America. John Wiley, New York, 272 pages
18. Billington, D.: The Tower and the Bridge – The New Art of Structural Engineering, 306 pages. Princeton University Press (1983)
19. ESPAS - Espas European Strategy and Policy Analysis System, "Global Trends to 2030: Can the EU meet the challenges ahead?", 80 pages (2015)
20. Fujino Y, Kimura K, Tanaka H (2012) Wind resistant design of bridges in Japan. Springer, 264 pages, 256 pages
21. Gazzola F (2015) Mathematical models for suspension bridges - nonlinear structural integrity. Springer International Publishing, Switzerland, 259 pages
22. Sheppard S, Tongue B (2006) Statics – analysis and design of systems in equilibrium. Wiley, pp 38–50
23. Steinmann DB (1922) A practical treatise on suspension bridges, their design, construction and erection. Wiley, 299 pages
24. Tezcan S, Arıoğlu E (1994) Çanakkale bridge. In: Proceedings of the third symposium on strait crossings, alesund, Norway, pp 21–36
25. Xu YL (2013) Wind effects on cable-supported bridges. Wiley, 742 pages

Bridge Failures and Mitigation Using Monitoring Technologies

Selcuk Bas and Necati Catbas

Abstract The current study aims to understand the reasons for bridge failures. It considers the recent damages and collapses of in-service real bridge structures and discusses mitigation methods based on monitoring technologies. Transportation systems serve a crucial function in the strategies for mitigating bridge damages and failures. After investigating recent damages and failures in real bridges, structural bridge failures are classified in this study according to their safety and operational function. For each function, global and local level failures are also defined with two groups: major/long-span and highway bridges. In the light of this classification, structural monitoring methods are identified according to global and local failures. Along with standard Structural Health Monitoring (SHM) systems with comprehensive sensor networks, at vision-based SHM system and developments in this area are shown to provide some opportunities for mitigation of bridge failures resulting from service loads rather than natural hazard-induced failures.

1 Problem Definition: Bridge Failures

Bridge structures are the main part of civil infrastructure systems in the transportation network of a country. Due to their complexity and the critical function they provide, special attention has been given to them from local and federal governmental bridge authorities. For reliable damage/failure estimation and mitigation approach, the structural response of bridges to service and extreme load cases must be accurately identified. According to ASCE's 2017 infrastructure report card, 10% of 614,387 bridges in the U.S. are fifty years or older. Among them, 56,007 bridges (9.1%) are defined as structurally deficient. The rehabilitation cost for these defected

S. Bas
Department of Civil Engineering, Bartin University, Bartin, Turkey
e-mail: sbas@bartin.edu.tr

S. Bas · N. Catbas (✉)
Civil, Environmental and Constructional Engineering Department, University of Central Florida, Orlando, USA
e-mail: catbas@ucf.edu

P. Gülkan et al. (eds.), *Developments in International Bridge Engineering*, Springer Tracts on Transportation and Traffic 17, https://doi.org/10.1007/978-3-030-59169-4_3

bridges is predicted to be 123$ billion. In addition to understanding the behavior of bridges under different loading and environmental conditions, the development of new effective assessment concepts and monitoring technologies enables bridge engineers/owners/authorities to mitigate bridge failures/collapses by reducing their structural deficiencies. Thus, it is essential to make more understandable the failures of recent in-service real bridges.

The key reasons for collapse of recent reinforced concrete and steel bridges are mostly based on deterioration in steel and concrete due to corrosion and other environmental effects, fatigue and poor or inadequate design considerations against unexpected loading and natural events. Specifically, shear failure and post-tensioning failure for concrete bridges and fatigue and buckling problems for steel bridges are the most commonly seen failures resulting in total collapse. For example, the well-known collapse of the Tacoma Narrows suspension bridge in the U.S. occurred because the dynamic effects of wind load was disregarded. Thus, a wind tunnel test was then implemented for special structures [1–8]. The following recent examples of bridge failures also reveal the importance of effective inspection, management, and use of structural monitoring systems in order to avoid possible bridge failures.

The I-35 Mississippi River Bridge in Minneapolis, Minnesota collapsed as depicted in Fig. 1, and 13 people died. After the catastrophic failure, an independent federal agency, the National Transportation Safety Board (NTSB) [9], conducted a detailed investigation and concluded that the stool plate of a diagonal member yielded under service load. Based on global failure of the bridge due to inadequate design of the gusset plate, some other independent researchers [8, 10, 11] underlined structural members of steel bridges designed in the U.S. during the 1960s should be reassessed in terms of dimensions and loading capacity with a proper maintenance

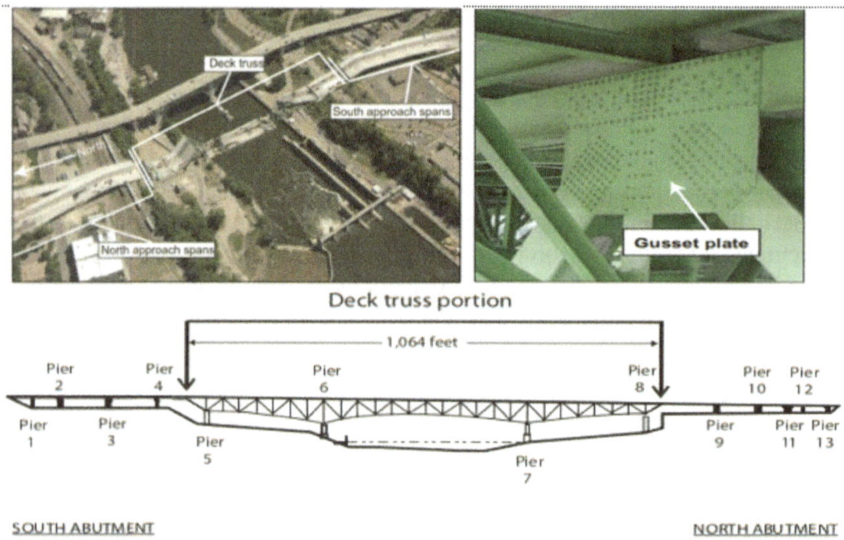

Fig. 1 Catastrophic failure and elevation layout of The I-35 Mississippi River Bridge [9]

and inspection strategy. Similar stool plate buckling failure was reported in 1996 for the deck truss bridge of the Grand Bridge near Cleveland, Ohio [8, 10, 11]. Thus, inadequate design considerations for dimensions and buckling strength are one of the main reasons for failure of steel bridges.

Antoher recent dramatic bridge failure was the collapse of the pedestiran bridge under construction at FIU (Florida International University) in Miami, Florida. This failure killed six people and injured eight people. It also crushed eight vehicles. According to the first preliminary and updated second report of NTSB with supplemented and corrected information, the bridge, as demonstrated in Fig. 2, was reported to have collapsed due to over estimation of sections of critical truss members (11 and 12), as shown in Fig. 2, and underestimation of demand load for these sections. This failure directly pertined to inadequate design considerations and was proved by observing the relative sag of the northern part of the bridge before collapse [12].

The Morandi Bridge in Italy, as shown in Fig. 3a, was severely damaged on August 14, 2018. In this severe damage, a noticeable part of the bridge, which corresponds to about 243 m of the viaduct over the Polcevera River in Genoa, was totally collapsed, and 43 people died at this event. The collapsed section of the bridge, as seen in Fig. 3b, was designed with a self-balancing methodology. Possible reasons for such heavy damage have pertained to the fatigue problem in the tendon elements and reduced connection between stay and transverse link due to deterioration [13].

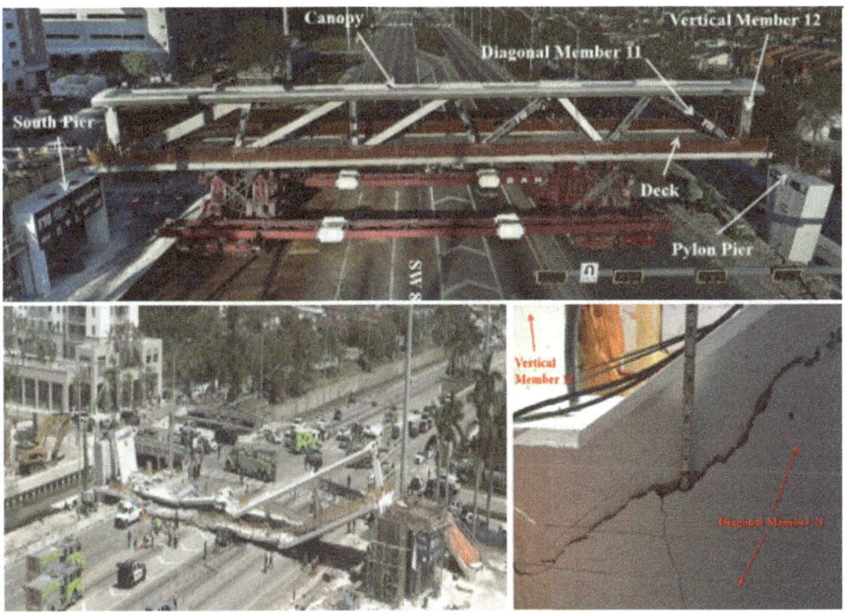

Fig. 2 Collapse of The FIU under construction pedestrian bridge [12]

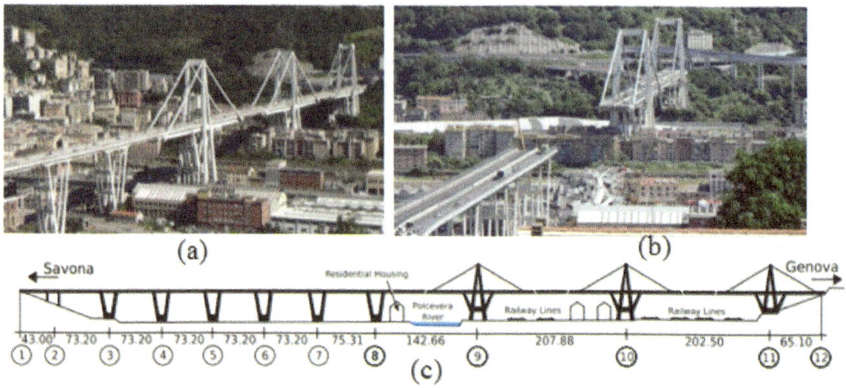

Fig. 3 The Morandi Bridge: (a) general view (b) collapsed case and (c) elevation layout [13]

Given the recent collapse mechanism of certain existing bridges, the need to develop effective assessment, maintenance, management, inspection, and safety strategies to mitigate damages, failures, and total collapse of bridges is remarkable. For this aim, one of the most reliable inspection approaches is the Structural Health Monitoring (SHM) system that enables engineers/bridge owners to measure the operating loading environment and critical response of a system in order to track and evaluate incidents, anomalies, damages, deterioration, etc. With the major leaps in sensing technology, a SHM system provides some important opportunities to make the inspection work of bridges easier and very quick. Considering the needs and some complementary new technologies (computer vision, video camera, IRT: infrared thermal camera, AI: artificial intelligence, LIDAR: 3-D point cloud) for a SHM system, it is clear that the failure problems of bridges will be reduced by SHM to some extent. Therefore, bridge failures and damages are classified in this study according to the aforementioned considerations of a SHM system.

As seen in Fig. 4, bridge failures are divided into safety/operational, major/highways bridge and global/local groups. Depending on the general definitions of the failure events, the physical parameters to be measured by a SHM system are specified. With the separation of the failure events into three groups of before, during and after, standard SHM system (wired/wireless) and vision-based (noncontact) SHM system (computer vision, video camera, IRT, AI, LIDAR) are assigned to the each group. The schematic representation can be adapted with reducing and extending list of the sensors and failure events, and others.

2 Mitigation of Bridge Failures with SHM Systems

As shown in Fig. 4, the general SHM approach is considered as standard SHM and vision-based SHM. Based on their response measuring capabilities and preferences,

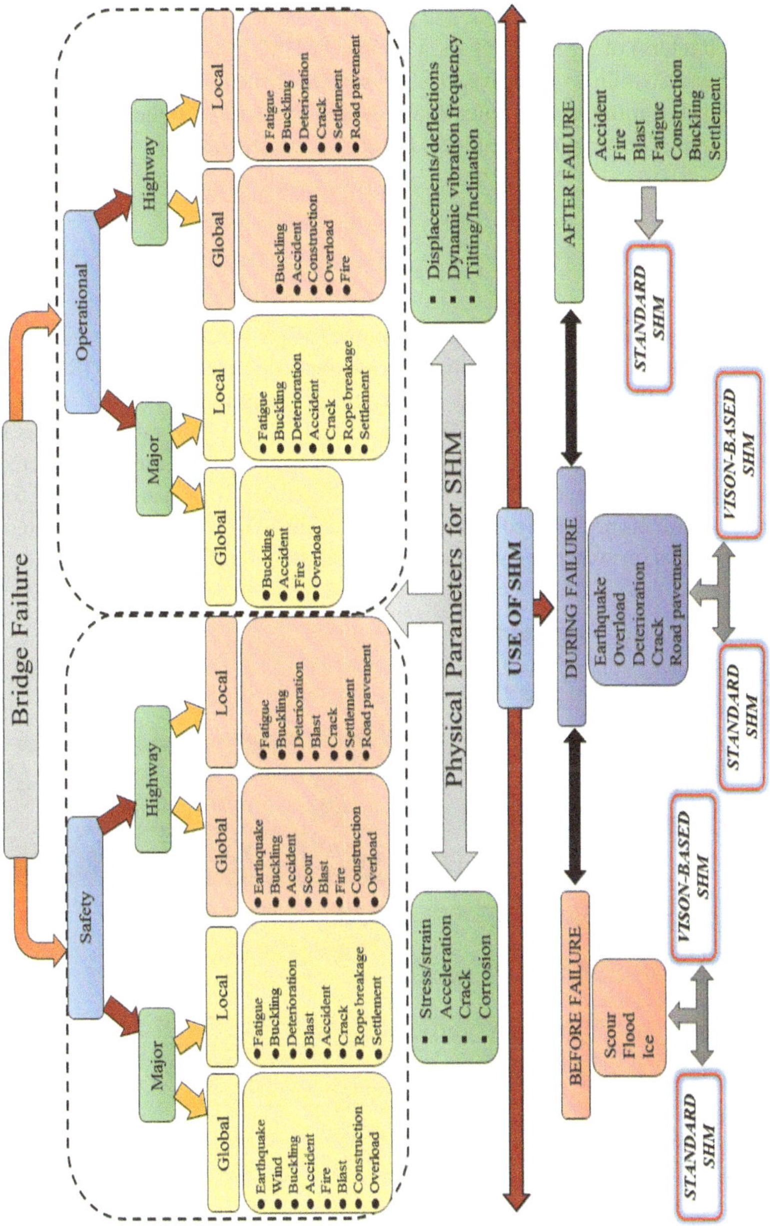

Fig. 4 Classification of bridge failures according to SHM system

the implementation of these are specified by defining the failure cases as before, during, and after. The possible failure events defined for global and local failure levels are also assigned to these cases. Considering these failure events, which are determined for failure cases and levels, and possible measuring outputs given in Fig. 4 (physical parameters) of a SHM system, a standard SHM system is seen to be used for all failure cases while a vision-based SHM system is seen to be utilized for only before and during failure cases. Thus, more details are given for the vision-based SHM system considering global and local bridge failure levels.

2.1 Local Level

For local level damage/failure estimation using the vision-based SHM method, one well-known example is the use of an infrared thermography camera (IRT). As revealed in Fig. 5, the developed deck top scanning integration system at CITRS (Civil Infrastructure Technologies for Resilience and Safety) Laboratory enables a user to find the location of damages and to classify damage types based on detected damages on bridge deck pavements. Artificial intelligence (AI) can also be used for local damage estimation. A mixed-reality method using AI to autonomously detect and quantify cracks on structures was developed, as shown in Fig. 6.

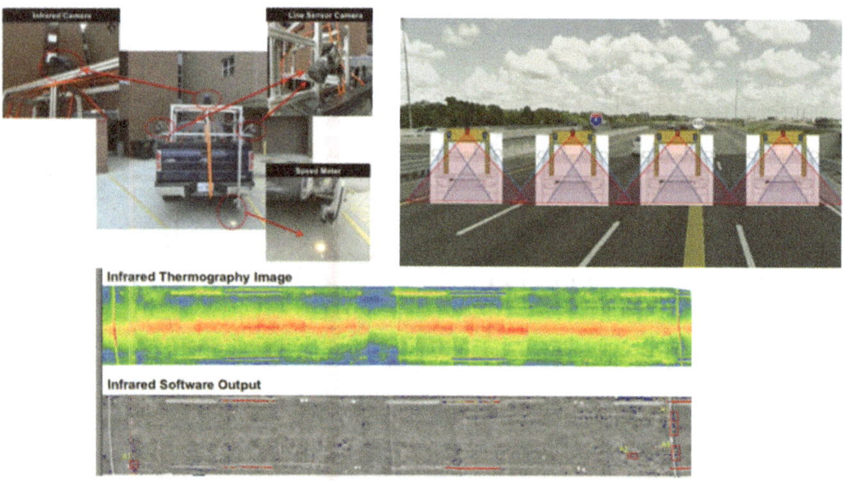

Fig. 5 The use of IRT for local damage mitigation [14]

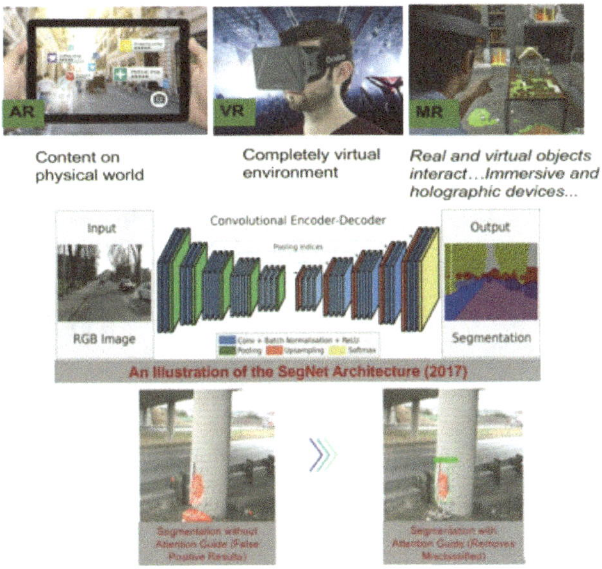

Fig. 6 The use of AI (mixed-reality) for local damage mitigation [15]

2.2 Global Level

The most effective and low-cost vision-based SHM approach for global failure estimation and mitigation is the use of the outputs of video camera data. As demonstrated in Fig. 7, this approach enables the prediction of overall bridge response with global response outputs, such as damage detection, load rating, load estimation, load distribution and structural displacement. By developing a statistical pattern recognition algorithm and using unit influence lines, global damage can be identified, localized, and quantified properly with video camera data, as depicted in Fig. 8.

The utilization of 3-D point cloud technology (LIDAR) is another effective approach to identify damages and to determine the performance of bridges. By using this technology to develop the current geometry of bridges after a loading event, as shown in Fig. 9, bridge failure and performance can be estimated with numerical analysis (finite element-FE).

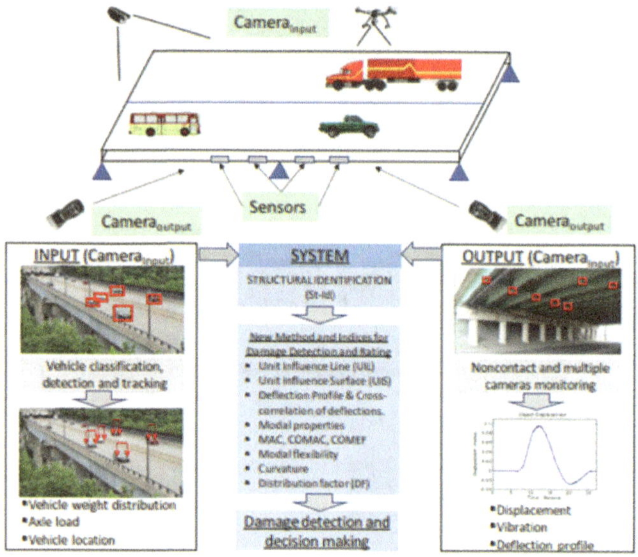

Fig. 7 Schematic presentation of the use of video camera for global damage estimation [16]

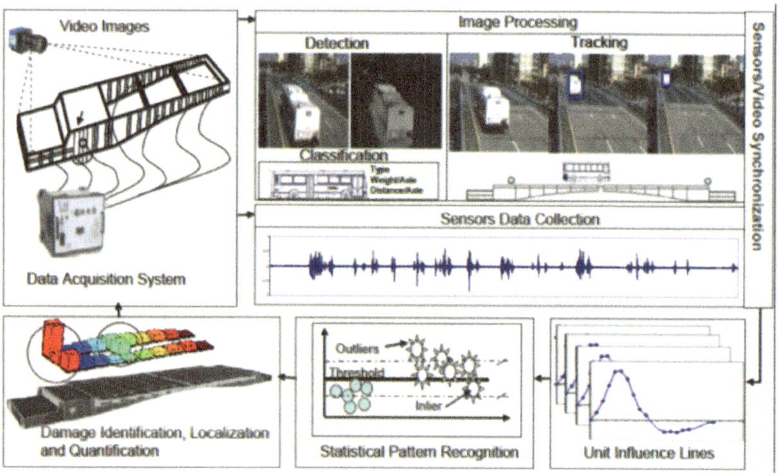

Fig. 8 Damage identification with the use of video camera [17]

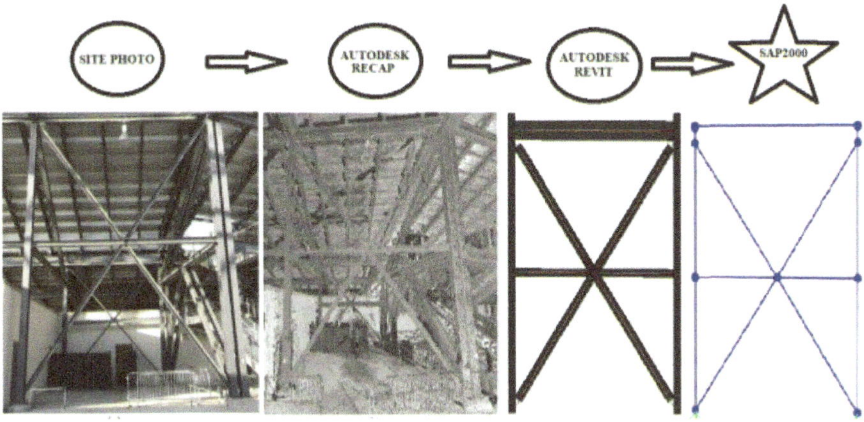

Fig. 9 Real model and developed model with LIDAR technology [18]

3 Concluding Remarks and Future Works

In this study, the recent important collapses of existing bridges are reviewed to discuss along with the promising SHM technologies that can be employed for monitoring, tracking, and maybe possibly predicting some of these failures. For this aim, bridge failures are divided into two parts according to functional limit states: safety and operational. Due to the fact that different failures occur on different bridge types, safety and operational bridge failures are classified by two bridge type groups: major (long-span suspension, cable-stayed bridges) and highway (standard bridges). For each bridge type, bridge failures are divided into two levels: global and local. Defining possible failure events for global and local failure levels and measuring parameters for the SHM system, standard SHM and vision-based SHM are assigned to three failure cases: before, during, and after. With the new developments in sensing technology and this classification, the following key points are obtained.

- Standard SHM and vision-based SHM systems have the potential to estimate and mitigate bridge failures resulting from flood, scour, and ice. Bridge failures are also identified before these failure events with these SHM systems.
- For the bridge failures that can occur during the failure events of earthquake, overload, deterioration, crack, and road pavement, standard SHM and vision-based SHM systems can be used.
- Bridge failures resulting from accident, fire, blast, fatigue, construction, and settlement can only be detected and mitigated with a standard SHM system.
- Rather than a standard SHM system, other complementary technologies (IRT, AI, and LIDAR) save on inspection time, have a lower labor cost, and allow utilization of data in infrastructure systems.
- Generally, SHM systems cannot fully mitigate naturally induced failure but can help identify partial failure.
- Decision-making should be integrated with a SHM system for more reliable results.

In the near future, opportunities with advances in camera technologies, non-contact, DIC, holographic lenses, IRT, AI, deep learning and LIDAR will open a new front for vision-based SHM systems. Thus, it is clear that bridge failures can be detected and mitigated with these developments.

Acknowledgements The authors kindly acknowledge all members of the Civil Infrastructure Technologies for Resilience and Safety (CITRS) research group at University of Central Florida for their prospective recommendations. Particularly, the authors highly appreciate Dr. Ricardo Zaurin, Dr. Shuhei Hiasa, Dr. Enes Karaaslan, Dr.Chuanzhi Dong, MSc. Jacob Cano for their endless supports and invaluable helps in the current study. The authors also thank Ms. Kaile'a Mosely of CITRS for editing the paper. The first author also wants to acknowledge the Scientific and Technological Council of Turkey (TUBITAK) through the grant number 2219. The second author would like to express his gratitude to National Science Foundation (CMMI 1463493), National Academies (NCHRP IDEA Program Project 222) and Nexco-West USA for the support of works some of which are presented in this paper.

References

1. Reissner H (1943) Oscillations of Suspension Bridges. T ASME 65:A23–A32
2. Billah KY, Scanlon RH (1991) Resonance, Tacoma Narrows Bridge failure, and undergraduate physics textbooks. Am J Phys 59(2):118–124
3. Biezma MV, Schanack F (2007) Collapse of steel bridges. J Perform Constr Facil 21(5):398–405
4. A°kesson B, (2008) Understanding Bridge Collapses. Taylor and Francis, London
5. Green D, Unruh WG (2006) The failure of the Tacoma Bridge: a physical model. Am J Phys 74(8):706–716
6. Larsen A (2000) Aerodynamics of the Tacoma Narrows Bridge: 60 years later. Struc Eng Intern 10(4):243–248
7. Petroski H (2009) Tacoma narrows bridges. Am Sci 97(3):103–107
8. Subramanian N (2008) I-35 W Mississippi river bridge failure—is it a wake-up call? Indian Concrete J 82(2):29–38
9. National Transportation Safety Board (NTSB) (2008) Collapse of I-35 W Highway Bridge, Minneapolis, Minnesota, August 1, 2007. Highway accident report NTSB/HAR-08/03.Washington, DC
10. Hao S (2010) I-35 W Bridge collapse. J Bridge Eng 15(5):608–614
11. Liao M, Okazaki T, Ballarini R, Schultz AE, Galambos T (2011) Nonlinear finite-element analysis of critical gusset plates in the I-35 W Bridge in Minnesota. J Struct Eng 137(1):59–68
12. National Transportation Safety Board (NTSB) (2018) Collapse of pedestrian bridge under construction, Miami, Florida (HWY18MH009).Washington, DC
13. Gian MC, Matteo M, Gerard JO, Nicola S, Ricardo M, Daniele M, Paolo MC, Rui P (2019) Once upon a time in Italy: the tale of the Morandi Bridge. Struct Eng Int 29(2):198–217
14. Hiasa S, Catbas FN, Matsumoto M, Mitani K (2017) Considerations and issues in the utilization of infrared thermography for concrete bridge inspection at normal driving speeds. J Bridge Eng 22(11):04017101
15. Karaaslan E, Bagci U, Catbas N (2019) Artificial intelligence assisted infrastructure assessment using mixed reality systems. Transportation Research Record
16. Catbas N, Dong CZ, Celik O, Khuc T (2018) A vision for vision-based technologies for bridge health monitoring. In: Maintenance, safety, risk, management and life-cycle performance of bridges - the 9th international conference on bridge maintenance, safety and management, IABMAS 2018. pp 54–62
17. Zaurin R, Khuc T, Catbas FN (2016) Hybrid Sensor-camera monitoring for damage detection: case study of a real bridge. J Bridge Eng 21(6):05016002
18. Cano JA (2019) Point cloud technology for analysis of existing structures. MSc thesis, Department of Civil, Environmental, and Construction Engineering, University of Central Florida, Orlando, FL, USA

Bridge Design

1915 Çanakkale Bridge – Meeting the Challenge

Inger Birgitte Kroon, Henrik Polk, and Kent Fuglsang

Abstract The 1915 Çanakkale Bridge is a world record 2023 m main span suspension bridge crossing the Çanakkale Strait in Turkey to connect Europe and Asia. Detailed design has been prepared in a challenging short period of 12 months. Significant design challenges for aerodynamics, ship collision, seismic and poor soil conditions have in combination with the very tight design and construction schedule significantly impacted alignment and articulation of the bridge. To accommodate the windy location, the bridge girder is designed as a twin-box steel girder for the purpose of a wind-resistant design. The dense ship traffic including the world largest ships and a massive grow rate for the strait has governed the design of the deep water tower foundations. The bridge situation in a seismically active region has further affected significant parts of the substructure design. Seismic hazard and difficult geological site conditions impose tremendous challenges to the bridge design. Finally, the tight construction schedule has conditioned the design solutions to accommodate fast track construction. The paper summarize the engineering approach, design decisions regarding alignment and articulation, design challenges and results obtained for the detailed design of the bridge.

1 Introduction

Turkey serves as an important crossing between Asia and Europe. Its rapid economic growth together with increased number of tourists, agriculture and transit transport has led to chronic traffic congestion.

The existing transportation network is unable to accommodate all demands arising from the traffic growth. To address these problems, the Turkish Government announced the Vision 2023 programme, which aims to increase the road, rail and sea transport capacities. Part of the programme is building a suspension bridge being the first bridge crossing the Dardanelles, in Turkey known as the Çanakkale

I. B. Kroon (✉) · H. Polk · K. Fuglsang
COWI A/S, Kongens Lyngby, Denmark
e-mail: ibk@cowi.com

P. Gülkan et al. (eds.), *Developments in International Bridge Engineering*, Springer Tracts on Transportation and Traffic 17, https://doi.org/10.1007/978-3-030-59169-4_4

Strait, which will improve the transportation network on the west side of the country, promote socio-economic growth and tourism.

In particular, the bridge crossing will provide an alternative route for European traffic to and from Izmir, Turkey's 3rd largest city, avoiding Istanbul. The bridge with its main span of 2023 m is going to be the suspension bridge with the longest main span in the world when it opens to traffic.

2 Location of the Bridge

The 1915 Çanakkale Bridge (in Turkish 1915 Çanakkale Köprüsü) is located 200 km southwest of Istanbul, spanning the Çanakkale Strait, which connects Lapseki District to the Gelibolu District (Gallipoli).

The strait forms a natural division between Europe and Asia, and it connects the Marmara Sea with the Aegean and Mediterranean seas, see Figs. 1 and 2.

The bridge is located in the central part of the 321 km long Kınalı - Balıkesir Highway, which will connect the O-3 and O-6 highways in East Thrace to the O-5 highway in Anatolia.

The 1915 Çanakkale Bridge is part of the programme, which expands earlier major transportation projects such as the Gebze-Orhangazi-Izmir Highway with the Osman Gazi Bridge.

The project is expected to increase capacity, improve traffic flow and ease the present and future congestion problems.

COWI has been involved with other major bridge projects in the area, as can be seen in Fig. 2, namely the Osman Gazi Bridge over Izmit Bay which opened in 2016 and the 3rd Bosporus Bridge named Yavuz Sultan Selim Bridge, which also opened in 2016. COWI was responsible for the detailed design of the Osman Gazi Bridge which has many similarities to the 1915 Çanakkale bridge.

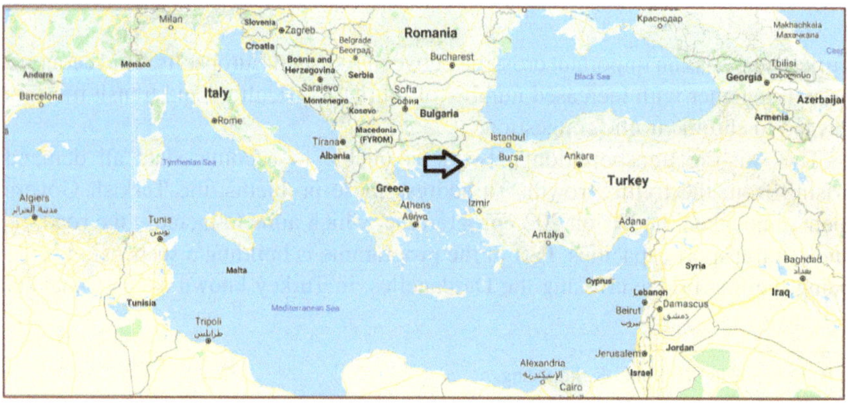

Fig. 1 Location of the bridge on the map. Source Google maps

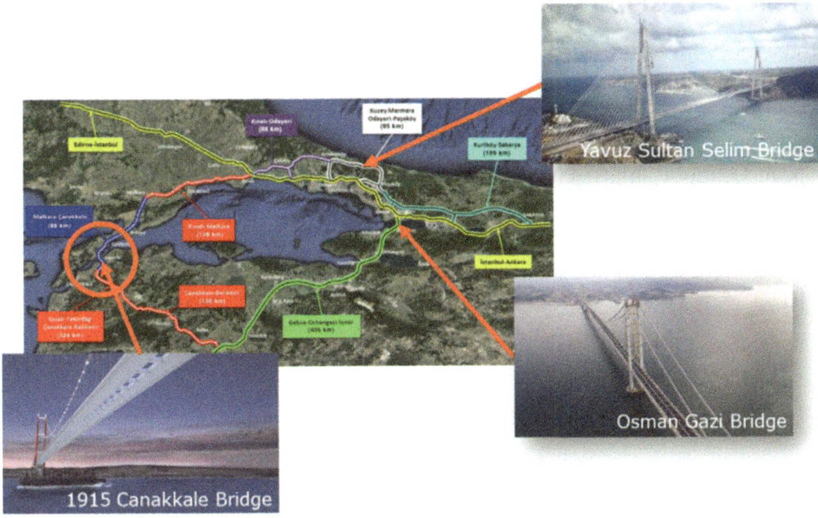

Fig. 2 Overview of the whole project. Source www.1915canakkale.com

3 Bridge Project Description

A consortium made up of South Korea´s Daelim and SK E&C and Turkey's Limak and Yapı Merkezi, named DLSY, won the tender for the construction of the bridge on 26th January 2017 by offering the shortest concession period of slightly more than 16 years, which includes a minimum revenue guarantee once the bridge is opened to traffic. The four companies each have an equal share of 25% in the project.

The General Directorate of Highways (KGM) awarded the 1915 Çanakkale Bridge and Highway Project within the framework of a Public Private Partnership model to the consortium, which subsequently established a joint-venture company (Commissioned Company), Çanakkale Otoyol ve Köprüsü İnşaat Yatırım ve İşletme A.Ş. (Çanakkale Motorway and Bridge Construction Investment Management Inc.).

The EPC contract for the bridge was signed in an official ceremony on 21st March 2017. The consortium will build, manage and operate the completed bridge for 16 years and two months under BOT basis. The bridge will be handed over to the Turkish Government after completion of the operation period. The bridge is expected to be built at a cost of around 2 billion USD. Another 0.8 billion USD covers the cost of the motorway.

The ground-breaking ceremony took place on 18th March 2017. Opening of the bridge is planned in 2022, in advance of the centenary anniversary of the declaration of the Republic of Turkey in 1923.

The Danish engineering consultant COWI was in 2017 awarded the design contract and COWI is responsible for the detailed design of the bridge.

4 General Design

The main structure will become the world longest suspension bridge with a main span of 2023 m and two side spans of 770 m each giving a total bridge length of 3563 m. Together with two approach viaducts the total bridge length becomes 4608 m. The European Gallipoli approach viaduct will be 365 m long and the Asian Lapseki approach viaduct will be 680 m long. An elevation of the bridge can be seen in Fig. 3 and a rendering of the suspension bridge can be seen in Fig. 4. The towers itself will have a height of 318 m and 334 m inclusive the special tower top, see furthermore Fig. 10.

4.1 Design Features and Challenges

Design of the suspension bridge system relies on sound FE modelling.

The global analysis model (GAM) is created in the structural design and analysis system IBDAS (integrated bridge design and analysis system) developed by COWI. The GAM is created as a parametric IBDAS FE-model using a combination of beam, shell and solid elements. Local models have a direct interface to the global model and are activated inside the global model obtaining easy loading and correct boundary conditions.

A rendering of the different geometrical elements within the IBDAS FE-model and element meshes can be seen in Fig. 5.

In excess of the general challenge related to design of a world record suspension bridge a number of specific challenges are being dealt with. First of all, the timeframe for design and construction is extremely tight. With a scheduled approximate 5 years

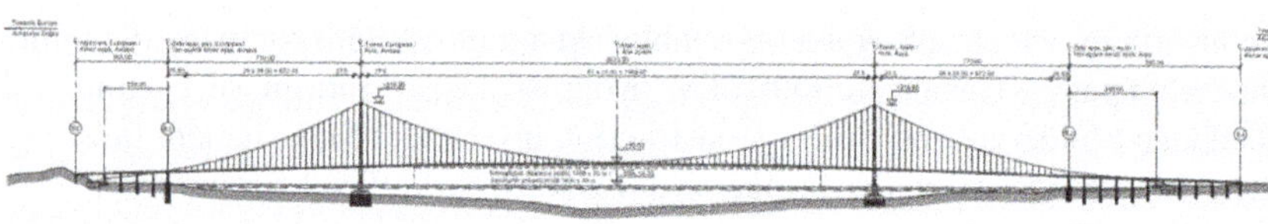

Fig. 3 Elevation of the total bridge

Fig. 4 Rendering of the suspension bridge

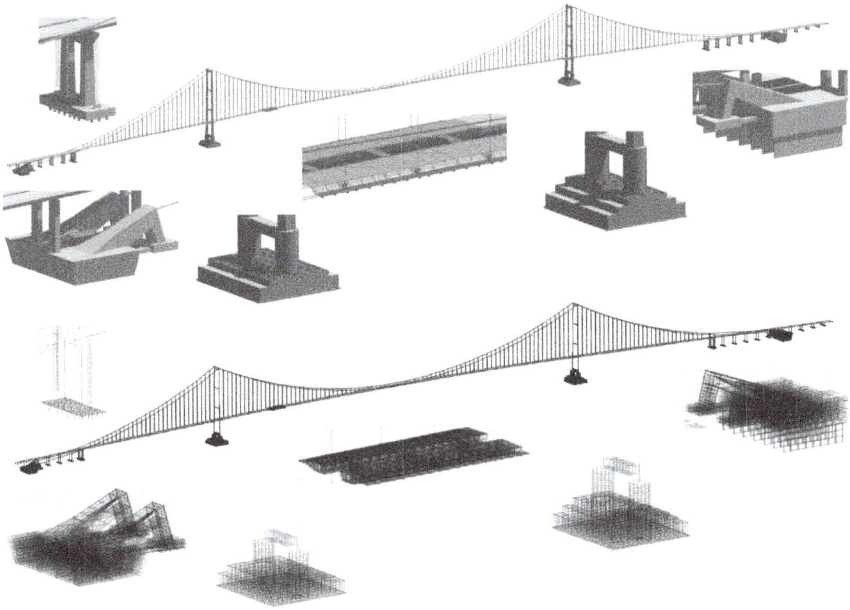

Fig. 5 IBDAS FE model – rendering and element mesh

Fig. 6 Caissons in flooded dry dock

for design, procurement and construction, the design process must be efficient and include optimised solutions to allow for fast construction.

Furthermore, there have also been a number of technical challenges specific to the 1915 Çanakkale Bridge which needed to be solved – among others related to the ground conditions, the seismic activity in the area, the intense ship traffic in the strait

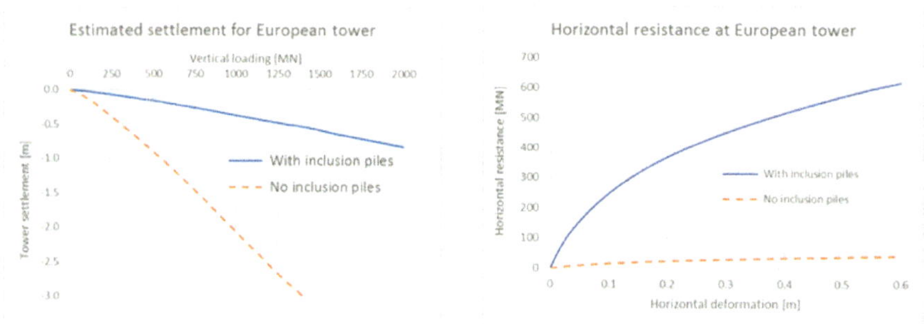

Fig. 7 Estimated settlements and horizontal resistance

of Çanakkale, the wind climate and the expected live load on the bridge. These have called for special focus on the seismic/earthquake design, aerodynamic design, ship collision risk analysis, live load modelling with focus on the special traffic on the bridge etc.

4.2 *Caisson Foundations and Towers*

The steel towers will be 334 m high, making the bridge the tallest bridge in Turkey. Each tower is founded on a cellular concrete caisson measuring approx. 83 m by 74 m in plan.

Each caisson was initially built in a dry dock that had been prepared on the European side of the strait. After the main cellular sections were constructed, the caissons were floated by controlled flooding of the dry dock and moved to deeper water called the wet dock (Fig. 6).

In the wet dock the concrete construction continued while the two caisson structures sunk deeper under the increased weight. When the concrete works were finished, two double walled 18 m diameter steel cylinders were installed on top of each caisson for later support of the tower legs. The 23-26 m high steel cylinders were moreover to ensure a controlled immersion of the caissons at the final position.

Simultaneous to the construction of the caissons, the seabed at the tower locations was being prepared for the placement of the caissons. At the seabed the ground consists of Holocene clay deposits at the European tower and Pleistocene clay and sand deposits at the Asian tower, followed by a Miocene mudstone formation below at both locations. The seabed was initially prepared by dredging to get a levelled plateau. Then 2.5 m diameter open ended steel inclusion piles were driven into the mudstone. 203 piles were installed at the European tower location with lengths up to 46 m and 165 piles at the Asian tower location with lengths of 21 m.

The inclusion piles reduce tower settlements (for the European tower by about 80%) and increase the lateral resistance of the foundation in the event of ship impact or seismic action (Fig. 7).

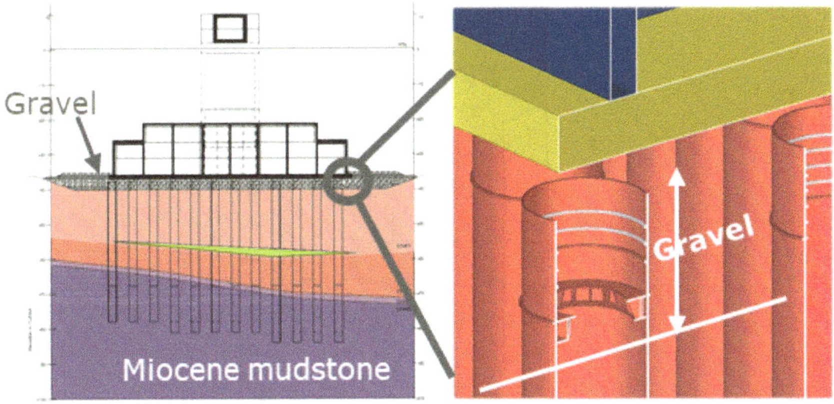

Fig. 8 Ground improvement and gravel bed for tower foundations

However, the piles are not directly connected to the caisson as can be seen in Fig. 8. For ensuring of the load transfer from caisson to steel piles a 3 m thick gravel bed is placed around the head of the piles on which the caisson is placed after the controlled immersion. This arrangement allows the tower/caisson to slide during a major seismic event.

After placement of the gravel bed on the seabed and finalisation of the works in the wet docks, the caissons were towed to the final position and sunk to the seabed which was at 37 m and 45 m below the water level at the European and Asian sides respectively.

After immersion of the caissons the steel shafts were filled with concrete between the double steel walls and solid 10 m high plinths were cast on top of each steel shaft. These plinths are interconnected with a tie-beam making the two towers legs acting as one unit particularly for ship impact- and seismic load situations.

The 318 m high towers are being constructed out of steel, primarily to allow for fast erection. They comprise tapered box sections with a chamfered corner for better dynamic performance in the wind. A plan of one tower leg can be seen in Fig. 9 with longitudinal flat stiffeners and cross frame.

The towers are divided into block elements which are prefabricated, 32 blocks in total for each tower leg and 3 blocks for each of the three cross beams. The size and weight of the blocks are primarily driven by the erection method. The first five very heavy blocks of each tower leg having a height of up to 11.0 m and a weight of up to 800 tons. They were designed to efficiently utilise the capacity of the floating crane that will erect these blocks one by one on top of each other.

The blocks are interconnected by horizontal block joints with welded skin plates and bolted splice connections of the internal longitudinal flat stiffeners.

For placement of the blocks above block 5 at level + 60 m, a heavy lifting tower crane will be utilised with a lifting capacity of 120 tons. This required the normal block elements to be subdivided into 2 panels, while 4 panels are requested for the

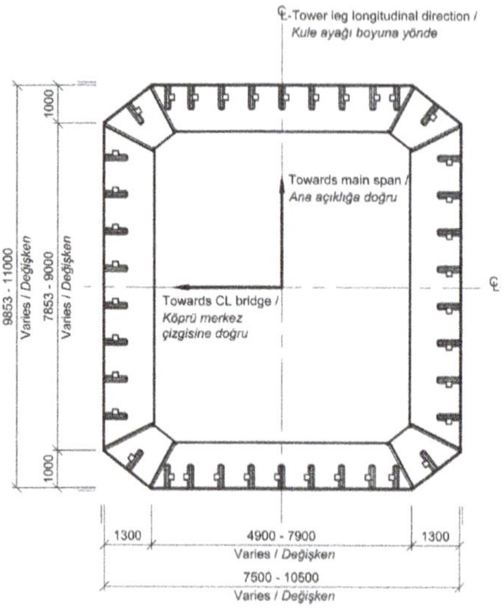

Fig. 9 Section of tower leg, cross frame spacing variable 2.2–3.0 m

Fig. 10 Renderings of the suspension bridge

special blocks at the cross beams and at deck level. The panels are interconnected by bolting at the corners of the block.

This method enables fast construction with the possibility to erect further blocks above before finalising the welding works.

4.3 Anchor Blocks, Side Span Piers and Main Cable

The ground conditions for the anchor blocks are critical in order to resist the massive design force of approx. 500MN from each main cable of the suspension bridge. The upper soil is, however, weak on both the European and Asian shorelines and the more competent Miocene formation is present at relatively large depths. It was therefore beneficial to move the anchor blocks away from the shoreline to where the Miocene formation daylights.

Fig. 11 Anchor block excavation, European

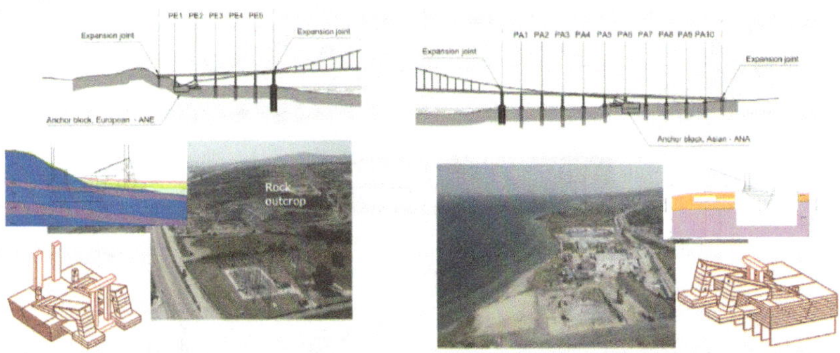

Fig. 12 Ground conditions and anchor blocks, European (left), Asian (right)

In order to get the anchor blocks placed on the Miocene formation the ratio of side span to main span was increased compared to usual and the cables were furthermore tied down at the side span piers. Increasing the ratio of side span to main span furthermore made it possible to place the side span piers at more favourable positions where the risk of ship impact is less and the geotechnical conditions are more favourable, i.e. on the shore on the European side and away from the underwater slope and into shallower water depth on the Asian side.

The anchor blocks are designed to minimise their height so the tension forces are transmitted as directly as possible into the foundation, reducing the over-turning

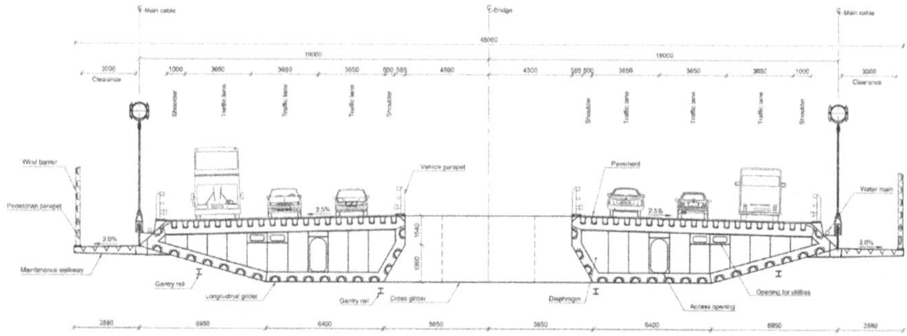

Fig. 13 Deck section – twin box girder with 9.0 m airgap

Fig. 14 Wind tunnel testing, tower model (Denmark), section model (Canada), bridge model (China)

Deck no	Deck configuration	a [°]	V_{cr} [m/s]	Comment
A1	9m gap, inner web 1 4m wind screens 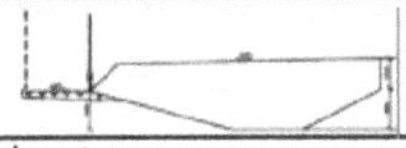	-1.5 0.0 +1.5	53 66 >71	Failed
B3	9m gap, inner web 2 5m wind screens 	-1.5 0.0 +1.5	64 >78 >69	OK
C3	9m gap, inner web 3 4m wind screens 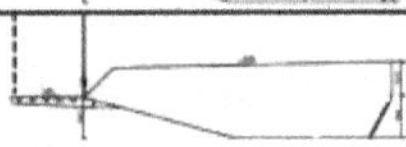	-1.5 0.0 +1.5	65 >77 >69	OK - preferred

Fig. 15 Testing of various profiles (Canada)

moment. The anchor blocks are positioned below the decks of the approach viaducts at each end meaning they have a setback of 250 and 350 m respectively from the end of the side spans of the suspension bridge (Fig. 11).

The main cables are formed of prefabricated parallel wire strand (PPWS) which are continuous between the anchor blocks. The setbacks of the anchor blocks from the side span piers implies that tie-down cables are introduced controlling the vertical position of the main cables at the bridge ends. Each tie down comprises four tensioned cables clamped to the main cable transferring loads directly to the piled foundation of the respective side span piers (Fig. 12).

4.4 Bridge Deck Structure

The bridge deck comprises two stiffened closed steel box girders spaced 9.0 m apart, connected by 3.0 m wide cross-girders every 24 m. The 9.0 m airgap between the two box girders ensures the aerodynamic stability of the bridge deck in strong wind for this world record free main span bridge. The overall width of the twin box girder becomes 45 m in total inclusive one maintenance walkway each outer side. The bridge will carry six lanes, three in each direction. The depth of the twin box girder is 3.5 m. The twin box girder reflects the advances in technology recently developed and also used on other very long span bridges (Fig. 13).

The underside of the twin box girder is located at level + 82.5 m at centre main span ensuring a navigation clearance for the ships of 70 × 1600 m.

The approach span sections at each end are of prestressed concrete box section construction.

4.5 *Special Analyses and Tests*

4.5.1 Aerodynamic Testing

Aerodynamic modelling and testing of long-span bridges are essential in understanding the dynamic response of the bridge structure due to wind loading and to optimise the design to ensure stability against the wind. Analysing of local wind data gave a basic 10 min mean wind speed of $v_{10} = 29$ m/s at level +10 m, giving $v_{10} =$ 46 m/s at maximum deck level of +86 m.

The wind tunnel testing, see Fig. 14, was carried out in three locations around the world, each investigating different specific elements and characteristics:

- Deck section model at 1:60, followed later by 1:30 scale in Canada
- Tower section model (1:80 scale), full tower model (1:225) and tower erection stages (1:225) in Denmark
- Full bridge model (1:190) and deck erection stages (1:190) in China

The aerodynamic stability of the bridge structure was verified through wind tunnel tests of the full aeroelastic model of the bridge. The twin box girder was verified for the selected airgap of 9.0 m and for the towers additional damping was introduced by an active mass damper (AMD) positioned 2/3 up in each tower leg.

Each box girder is shaped to minimise the effects of wind forces and to maintain aerodynamic stability. A number of different box profiles were tested to optimise the behaviour with variations to the geometry of the inner web, gap width and use of different heights of outer wind screens (Fig. 15).

The flutter stability is dependent on mean twist angle of the bridge girder with its nose up due to wind loading and to ensure stability, the models must prove that the deck remains stable for a wind speed up to 69 m/s.

4.5.2 Ship Impact

The Çanakkale Strait connects the Mediterranean and the Black sea through the Sea of Marmara and constitute an important link for freight transport with approximately 44,000 registered vessels passing the strait in 2016. The traffic today includes 260,000 DWT bulk carriers, 167,000 DWT oil tankers and 16,000 TEU container ships. Even though these are not the largest ships in the world fleet they are of very significant size and it is considered that there may be even larger ships in Çanakkale Strait in the future.

The navigation clearance envelope for the bridge is 1600 m wide by 70 m high, centred at the main span. This will allow for safe passage of the approximately 44,000 ships annually as of today and as well account for a potential large future increase in ship traffic. However, even with this large navigation span and appropriate traffic control, the caissons foundations for the tower structures need to consider potential

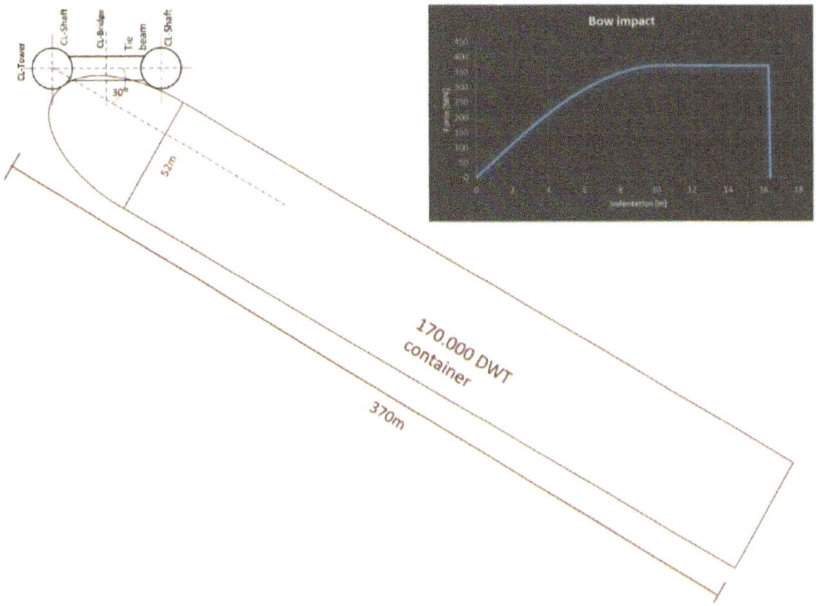

Fig. 16 Potential design vessel

impact from ships. Also, the lower sections of the towers, up to 29.5 m above sea level, are exposed to ship impact.

The ship impact design requirement is based on ship collision risk analysis leading to required resistance for a ship, impacting at an angle of up to 30 degrees with a global impact force of 370 MN. A potential design vessel is shown in Fig. 16 to indicate the size of the vessel compared to the dimensions of the composite caisson shafts. These loads are then transmitted primarily down through the composite shafts and caissons to the soil-improved seabed.

Semi-local and local impact govern the steel tower leg design up to + 29.5 m. The box sections are stiffened with horizontal diaphragms and skin plate thickness has been increased to deal with these actions. Horizontal stiffeners are also added to increase local bending resistance of the skin plates.

Strict design criteria have been introduced to achieve minimal damage of non-accessible parts of foundations under accidental load. This has resulted in specific design challenges and detailing of reinforcement, e.g. diagonal bars in some walls and slabs (Fig. 17).

4.5.3 Seismic Analysis

The bridge is located relatively close to the North Anatolian fault, but does not cross it directly. The design criteria consider 3 potential events:

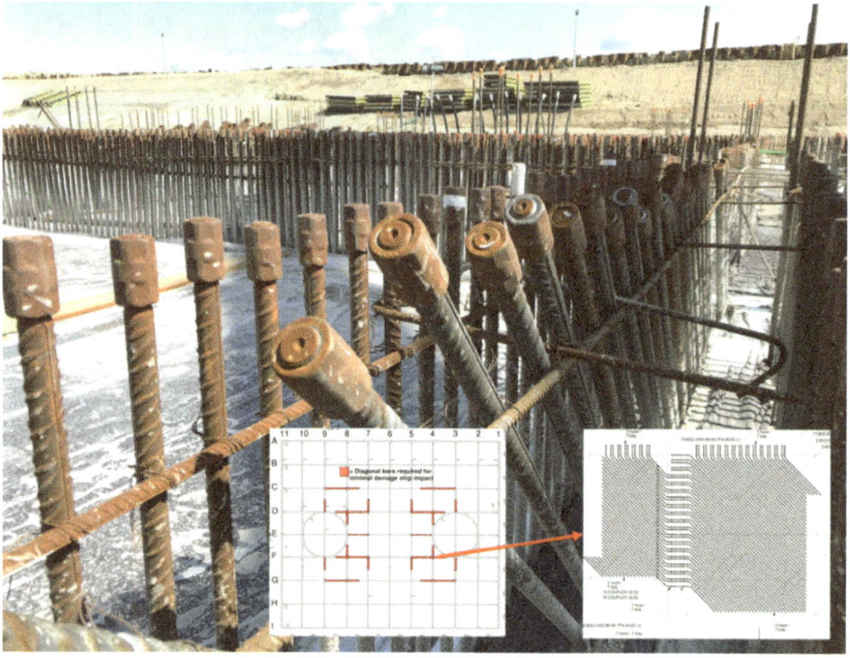

Fig. 17 Caisson - diagonal reinforcement of walls to resist ship impact

- Functional evaluation earthquake (FEE) – 145 year return period
- Safety evaluation earthquake (SEE) – 975 year return period
- No collapse earthquake (NCE) – 2475 year return period

To analyse the non-linear element behaviour of the bridge during a seismic event, dynamic analyses have been performed using the global finite element model and applying earthquake displacements in three directions at the six main supports, being two anchor blocks, two end supports and two towers. In total seven sets of time-displacement actions, which are to model seven different earthquakes, are applied for each of the three design earthquakes magnitudes/return periods. The analyses investigated the influence of using hydraulic buffers at the towers, of wind bearings and of soil-structure interaction.

4.5.4 Traffic Load on the Bridge

The traffic load on a bridge with a main span length of more than 2000 m becomes very important as the Eurocode system only considers bridges up to 200 m lengths. For bridges above this length the load models are therefore conservative and can instead be defined for the individual project. For the Çanakkale Bridge the Swedish national annex has been utilised for loaded lengths above 200 m.

The design therefore considers traffic loading as follows:

- Loaded lengths ≤ 200 m, the uniform distributed load (UDL) becomes 81.8 kN/m (2 × 3 lanes) based on Eurocode 1991–2 load model 1, 2, 3
- Loaded lengths > 200 m, UDL = 58.8 kN/m (2 × 3 lanes) EN 1991–2 SE-NA taking effect of long loaded length into account

Acknowledgements This paper is prepared based on the 2018 iBridge presentation "1915 Çanakkale Bridge design – meeting the challenge" presented by COWI on the Istanbul Bridge Conference, 5–6 November 2018.

The paper takes basis in the detailed design undertaken by the COWI project team and we are thankful to our colleagues who provided the expertise to the detailed design of the 1915 Çanakkale Bridge.

Further, we have to express our appreciation to the contractor DLSY (Daelim, SK E&C, Limak and Yapı Merkezi) for fruitful collaboration throughout the detailed design allowing for a to the point design of this world record span suspension bridge.

“Piccoli Angeli” Bridge Over Gorzone Canal in Cavarzere (VE)

Alessandro Stocco and Enzo Siviero

Abstract The “Piccoli Angeli” bridge was implemented by Zara Metalmeccanica Srl and designed by Prof. Enzo Siviero’s team. The bridge, with a 56 m span, is the synthesis of a design path that considered points of view related to culture and landscape as well as technical and economic aspects. The bridge, which has a strategic function in the countryside network context, memorializes through its name (Ponte degli Angeli) a tragic event that marked the history and now belongs to the memory of the place. The colours of the bridge (a dark gray for the scaffolding and white for the lateral bands) make it visible against the landscape, while the tapered and tense shape recalls the light curvatures that appear on the horizon. The result is a sober, elegant and economical work: the bridge was built for about 800,000 Euro.

1 Introduction

The “Piccoli Angeli” bridge was designed by Prof. Enzo Siviero’s studio in a long collaboration with the local administration. The contract for the bridge was awarded in 2015. After a year of work only the foundations had been constructed. This delay and several vicissitudes led to the dissolution of the contract with the first construction company. Zara Metalmeccanica Srl, the company which placed second in the original contract bidding, took over the contract, launching the metallic structure after only 5 weeks. From that point, the work proceeded expeditiously until the end of construction and the following subsequent inauguration and public opening of the bridge (Figs. 1 and 2).

A. Stocco (✉)
Nova Goriza University, IP Associates, Padova, Studio, Italy
e-mail: a.stocco@studioip.it

E. Siviero (✉)
eCampus University Rector, Padova, Italy
e-mail: enzo.siviero@esap.it

P. Gülkan et al. (eds.), *Developments in International Bridge Engineering*, Springer Tracts on Transportation and Traffic 17, https://doi.org/10.1007/978-3-030-59169-4_5

Fig. 1 Photo of the new bridge just after construction

Fig. 2 Image of the collapsed wooden bridge

2 The Reason for the Intervention

The new bridge replaces a concrete crossing built during the 1950's after the collapse of a wooden bridge that connected the two banks of the canal. 15 children, who were crossing the bridge as part of a parade for Children's Day, lost their lives in the collapse of the wooden bridge. The event left a mark on the territory, and the young victims are memorialized every year.

Archival research has brought to light the original designs for the concrete bridge, drafted in August 1950 (Fig. 3).

That replacement bridge - still in existence but closed to traffic and marked for demolition - has a Gerber type static layout and covers, with three spans, a total distance of 56 m between the abutments. The two shore-side spans are 15.5 m, with cantilevers of 8 m in the central span, while the central Gerber beam measures about 9 m (Fig. 4).

As early as 2007 the results of a preliminary static analysis highlighted problems with both the accessory elements of the bridge - the parapets and sidewalks - and the

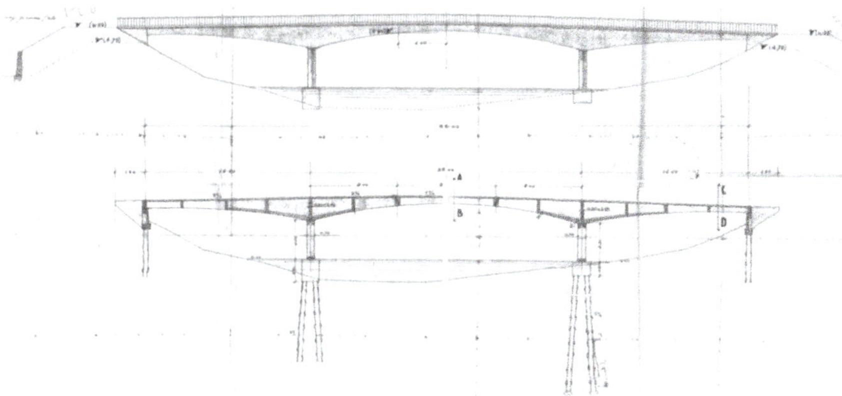

Fig. 3 Elevation view and section of the old concrete bridge, original design

Fig. 4 Photo of the bridge to be replaced before the start of work

load-bearing structure itself. Even at that time the bridge showed "evident signs of age, particularly at the pier sections".

Successive analyses showed a natural increase in points of deterioration in the bridge, both in its general character and the load-bearing structures themselves. This is particularly true regarding the areas of the intrados of the deck that increasingly show in more and more points, the progress of corrosion on the many reinforcing bars, which are now without reinforcement covering and have been deteriorated by exposure to atmospheric agents.

3 Design Criteria for the New Bridge

Because of the need to respect the Administration's budget, no particularly innovative technologies were put into place. We have chosen a thoughtful use of *state-of-the-art* knowledge related to the construction of bridges with reinforced concrete slabs, in order to optimize construction and maintenance costs (Fig. 5).

Therefore a study was carried out for a structure with one span utilizing the layout of a beam on fixed supports with encapsulated neoprene, with clear savings in the supports and joints (Fig. 6).

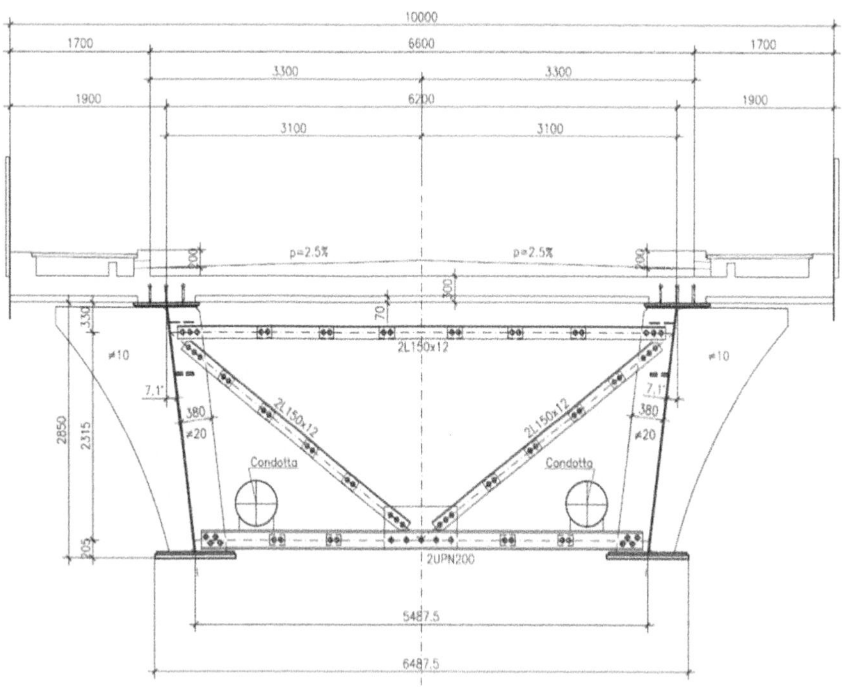

Fig. 5 Support diaphragm and bridge bearings

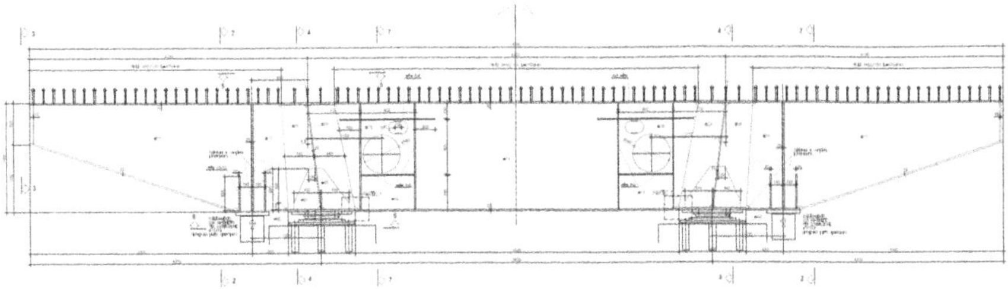

Fig. 6 Section at centreline

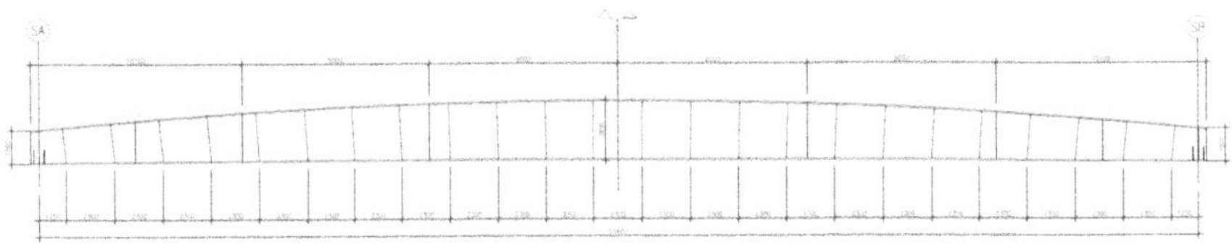

Fig. 7 Elevation of the longitudinal girders

The geometry of the longitudinal girders is the result of a study aimed at optimizing the static behaviour of the bridge with a consequent reduction of the quantity of steel and relative operations. Also studied was the possibility of incorporating the utility pipes between the longitudinal girders, which instead were positioned externally on the deck of the bridge to be replaced (Fig. 7).

With an eye towards aesthetics, chromatism and several fundamental details were studied for the successful insertion of the bridge into the landscape [1].

The dark colour of the steel girders, together with their slight inclination from the vertical plane and the cantilever slab, allows the light to have a soft effect, avoiding strong contrasts between light zones and shaded zones.

The side cantilevers (the transverse overhang brackets), which serve to stiffen the longitudinal girders, are radially arranged and their spacing has been studied so that they coincide with the parapet module, resulting in an image that is both geometric and dynamic.

Furthermore, because of the geometry of the single span, this solution has augmented the capacity of the hydraulic section if compared with that of the current bridge. Not having piers in the river, the new bridge in no way impedes the water flow of the canal (Figs. 8 and 9).

4 Design

The new bridge stands immediately upstream of the existing bridge, which will be demolished soon. The plan calls for wider driving lanes than the existing bridge,

Fig. 8 Side cantilevers

Fig. 9 Lateral view of the new bridge

especially at the intersections, in order to allow vehicles to handle the turning more easily.

The total length is 55.4 m and the width is 10 m (1.7 m for the sidewalks and 6.4 m for the carriageway).

The central structural body, consisting of two principal steel girders, is placed behind the deck slab extension, towards the centre. The cross section in the middle of the span is higher than the cross section at the sides, forming a longitudinal profile, in the extrados, shaped as a very stretched arch. This arrangement gives the bridge a fluid, slender aspect which, together with the external stiffening ribs aligned with the parapet posts, generates an interesting rhythmic play between light and shade [3] (Fig. 10).

The absence of masking elements and visual sophistications allows a clear reading of the structural behaviour of the bridge in relation to the distribution of force and

Fig. 10 Rendering of the new bridge design

pressure. The optimizations of the structure, in relation to typology and the forms in play, permit the attainment of a high level of economy and a stimulating aesthetic effect.

5 Structure

As anticipated, the static plan adopted consists of one beam on a fixed support at the extremities of the foundation beams. The structure of the bridge was developed in such a way as to cover the 56 m between river banks with a single span. It consists of two principal longitudinal steel girders - the height of which varies from 1.25 m (at the ends of the bridge) to 2.65 m (in the centre) - transversely connected by diaphragms. The side cantilevers serve to stiffen the longitudinal girders.

The deck beams are made solid by a 25 cm thick concrete slab which serves to increase the rigidity of the load-bearing system, as well as constituting the transversal element of support for applied loads. The slab is composed of a self-supporting 5 cm thick *predalles* plate (a kind of concrete stay-in-place formwork), on top of which a 20 cm thick supplementary casting is placed. A 5 cm thick binder and 3 cm thick asphalt concrete are successively placed.

The foundation structures are deep type, each consisting of 5 drilled bits 100 cm in diameter and 23 m long connected to foundation beams, upon which the longitudinal steel girders rest on elastomeric bridge bearings.

6 Treatment of Lateral Beam Surfaces

Regarding the treatment of external surfaces, the design calls for the use of high protection paints through a painting cycle carried out at the factory and consisting of preventative sandblasting with grade SA 2.5, an inorganic zinc primer min. 75 μ, an intermediate epoxy layer min. 80 μ, and a layer of polyurethane finishing min. 50 μ.

The chromatisms of the metallic structure were selected through a study of the surrounding landscape, which identified a particular range of tints. RAL 7022 "shadow gray" was then chosen for the longitudinal girders and the transverse deck overhang brackets, RAL 7023 "concrete gray" for the parapets and RAL 9016 "traffic white" for the lateral cladding.

7 Steel Structure Launching

The assembly of the new metallic deck was executed, after applicable checks, on the deck of the existing bridge, where the single segments were placed and then full-penetration welded.

The segment assembly operations took place alongside the insertion of two pipelines (500 mm in external diameter) that will be used for the adaption of a future water network.

The launch of the steel superstructure, 56 m long and about 200 tons in total weight, was carried out by synchronizing the work of three mobile cranes, the heaviest of which, with a 450 ton capacity, was positioned on the Boscochiaro side while the other two, each with 200 ton capacities, were positioned on the opposite side (the Martinelle side) (Figs. 11, 12 and 13).

Fig. 11 The steel superstructure in the assembly phase and utility water pipes between the longitudinal girders

Fig. 12 The new steel superstructure at the end of segment assembly

Fig. 13 Launching of the steel superstructure

8 Conclusions

The "Piccoli Angeli" bridge is a strategic component of the connection network, which, through its name and colours, connects to the history of the resident population and integrates itself culturally into its context [2]. Its geometries, studied to optimize the static bearing, have been calibrated with the goal of rendering this new work part of the surrounding landscape. After the initial problems that greatly slowed the work, Zara Metalmeccanica has completed the project in the predicted time with a budget of about 800,000 euro. The synergistic work between the administration, designers, and builders has produced a sober, elegant, and economical bridge.

Acknowledgements The cultural approach to this specific theme as the concept design of the work is the result of a synergic process developed by Alessandro Stocco and Prof. Enzo Siviero. The final design, together with the implementation of the work, has been completed thanks to the coordinated work of an interdisciplinary team. In particular, this paper contains a synthesis of the important analytical work developed by Pietro Zara, Mario Organte and his office, Federico Zago,

Luigi Rebonato, Michele Culatti, Gabriele Maragotto, Marco Zanon, and mayor Henri Tommasi together with Fausto Sanguanini (for the local administration).This work encompasses all stages of the construction, from the static and dynamic behaviour of the bridge to the transportation and implementation phases of the bridge.

References

1. AA.VV (2002) Infrastrutture e paesaggi contemporanei, edited by Maffioletti S., Rocchetto S.,Il Poligrafo, Padova, Italia
2. Arici M, Siviero E (2005) Nuovi orientamenti per la progettazione dei ponti e viadotti. Dario Flaccovio Editore, Italia
3. Culatti M, Siviero E (2005) Il ponte concettuale: il punto di partenza per una riflessione sul rapporto con il contesto, Il ponte concettuale, pp 65–68

Predicting Time-Dependent Deformations of Prestressed Concrete Girders

Levent Isbiliroglu, Robert W. Barnes, and David M. Mante

Abstract Camber prediction plays an important role in the construction planning of prestressed concrete girders. An accurate prediction is essential to avert unexpected cost and construction delays. The purpose herein is to compare the accuracy of the creep and shrinkage models such as AASHTO LRFD, ACI 209R-92, and *fib* MC 2010 on strain, curvature, and camber predictions. The experimental data were gathered from fifty-one full-scale girders including AASHTO BT-54, BT-72, and Type I and T-Beams. Compressive strengths at 28 days varied between 43 and 94 MPa. The ALCAMBER v1.0 software, utilizing a time-steps approach, was developed for the predictions and made available online. The results demonstrated that the camber growth and the growth of prestress losses were not accurately predicted and were overpredicted on average. The growth predictions up to the girder erection were most accurately predicted in terms of (i) prestress losses by AASHTO LRFD with an error of 68% and (ii) camber growth by ACI 209 with an error of 51%.

1 Introduction

Camber is the time-dependent upward deflection of a prestressed girder. Initial camber (i.e., camber at transfer) can be observed as soon as the prestressed strands are released (i.e., the prestress transfer) and occurs due to the elastic deformation of concrete, which is sketched as the superposition of dead load and prestressing forces in Fig. 1. It is also affected by steel relaxation and temperature variations prior to the prestress transfer. Long-term camber comprises the initial camber plus the *growth*

L. Isbiliroglu (✉)
ALTEN SA, Boulogne-Billancourt, France
e-mail: levent.isbiliroglu@alten.com

R. W. Barnes
Department of Civil Engineering, Auburn University, Alabama, Auburn, USA
e-mail: barnerw@auburn.edu

D. M. Mante
Department of Civil and Environmental Engineering, Lafayette College, Easton, Pennsylvania, USA
e-mail: manted@lafayette.edu

P. Gülkan et al. (eds.), *Developments in International Bridge Engineering*, Springer Tracts on Transportation and Traffic 17, https://doi.org/10.1007/978-3-030-59169-4_6

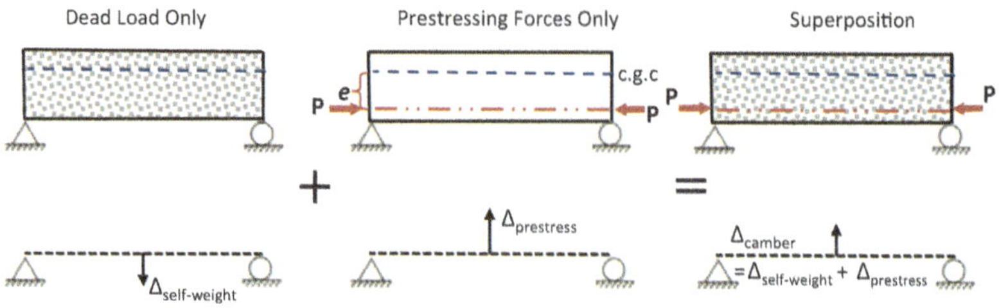

Fig. 1 Illustration of camber deflection. *c.g.c: center of gravity of concrete section

of deformations (after prestress transfer) during a girder's service life. It is affected by the interaction of prestressed steel relaxation, creep and shrinkage of concrete, maturing concrete properties, and ambient conditions.

The prediction of camber is important during the construction planning of prestressed concrete girders to avoid unexpected cost and construction delays. Underpredicted camber may require altering the deck forms during construction in order to achieve the desired deck elevation. Overpredicted camber may require increasing the quantity of deck concrete and cause bridge sagging under superimposed dead load (i.e., due to the deck and the barrier rails).

Camber at prestress transfer can be estimated by using a classical beam-deflection method and linear-elastic material properties. Predicting camber *growth* necessitates a more complicated approach. There has been a variety of approaches such as the simple PCI multiplier method, the approximate time-steps method, the axial strain and curvature method, the prestress loss method, and the incremental time-steps method, which is the most complex approach [1, 2]. The use of theoretically accurate numerical methods is favored over the simple methods [1, 3, 4].

For accurate camber predictions, it is recommended to use the material properties representative of local material properties and to include the effects of storage and construction practices [2, 5–8], but most of this information is not available to engineers during the design. Also, there has not been a consensus in the literature on which material prediction model(s) to use for most accurate camber *growth* predictions.

This study questions the effect of the creep and shrinkage models such as AASHTO LRFD [9], ACI 209 [10], and *fib* MC 2010 [11] on the prediction of camber along with strain and curvature at midspan with the representative material properties from the experiments. The experimental data cover fifty-one full-scale bridge girders. The observed field data is then compared with the data predicted by ALCAMBER v1.0, which can be downloaded by completing the inquiry form available at the website of Highway Research Center (www.eng.auburn.edu/research/centers/hrc/hrc-info-pages/research/software.html).

2 Methodology

The ALCAMBER software, utilizing the incremental time-steps method for non-cracked prestressed concrete sections, was developed for this research. It integrates the deformations caused by girder's self-weight, prestressing forces, creep and shrinkage of concrete, and relaxation of prestressed steel.

The implemented algorithm in Fig. 2 computes the incremental curvatures and axial strains at each cross section in each time step. It includes three fundamental assumptions: 'a plane section remains plane', the linear-elastic stress-strain behavior for stress-induced strains, and equilibrium of internal stresses with external forces [5–7].

The initial calculations in Fig. 2a consider the elastic shortening of concrete and the steel relaxation that occur at and before prestress transfer. It calculates the initial strains and curvatures. Time arrays are defined with a nonlinear function to better characterize the rapidly changing earlier behavior. The array for the modulus of elasticity (MOE) development is also created. A new MOE model was developed as given in Eq. 1. It adapts the time development function in *fib* MC 2010 including a coefficient related to strength class of cement. The coefficient is back-calculated with the use of the supplied MOE values at transfer, $t_{transfer}$, and 28 days:

$$E_c(t) = \exp\left\{ \frac{\ln\left[E_c\left(t_{transfer}\right)\right]}{1 - \sqrt{28/t_{transfer}}} \times \left[1 - \sqrt{^{28}/_{t}}\right] \right\} \times E_{c,28} \quad (1)$$

Where $E_c(t)$: modulus of elasticity at an age of t in [days]
$E_{c,28}$: modulus of elasticity at an age of 28 days
t: age of concrete in [days]

In the step in Fig. 2b, steel relaxation and unrestrained incremental time-dependent deformations due to creep and shrinkage of concrete are calculated for each cross section and each time step. Over the girder depth, the shrinkage deformations are considered to be constant; in contrast, the creep deformations are considered to be directly proportional to the cross-sectional stress distribution. The steel relaxation is modeled with the common expression [5]. Three creep and shrinkage prediction models are implemented [7] according to AASHTO LRFD, ACI 209 and *fib* MC 2010.

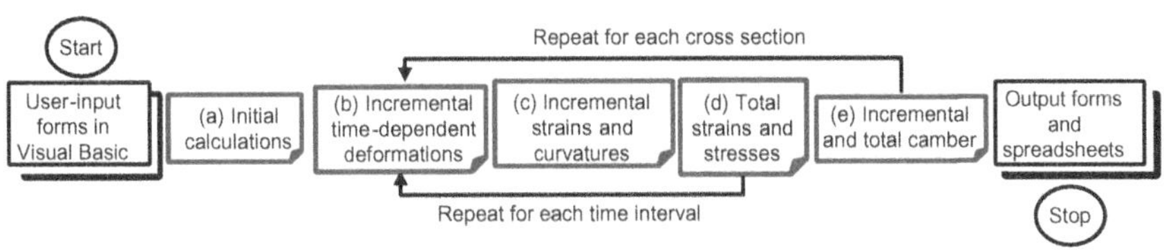

Fig. 2 Algorithm of ALCAMBER v1.0 developed at Auburn University

In the step in Fig. 2c, the time-dependent deformations and the maturing concrete properties are combined to compute the incremental axial strain and curvature at each section in each time step. The effects of restraint on creep and shrinkage (due to the steel presence) are considered, and the above-mentioned assumptions are satisfied by Eqs. 2 and 3 [5]:

$$\Delta\varepsilon_{cen} = \frac{A_c}{A_{tr}}\left(\Delta\varepsilon_{cen.cr} + \Delta\varepsilon_{c,sh}\right) - \frac{\Delta\phi_{c,cr}\left(n_p \sum A_{p,k} y_{p.k} + n_s \sum A_{s,k} y_{s,k}\right)}{A_{tr}} - \frac{\frac{1}{E_c}\sum \Delta f_{p,R,k} A_{p,k}}{A_{tr}} \tag{2}$$

$$\Delta\phi = \Delta\phi_{c,cr}\left(1 - \frac{n_p \sum A_{p,k} y_{p,k}^2 + n_s \sum A_{s,k} y_{s,k}^2}{I_{tr}}\right) - \frac{\frac{1}{E_p} n_p \sum \Delta f_{p,R,k} A_{p,k} y_{p,k}}{I_{tr}} - \frac{\left(\Delta\varepsilon_{cen.cr} + \Delta\varepsilon_{c,sh}\right)\left(n_p \sum A_{p,k} y_{p.k} + n_s \sum A_{s,k} y_{s,k}\right)}{I_{tr}} \tag{3}$$

Where $\Delta\varepsilon_{cen}$: incremental strain in concrete at the centroid of transformed (tr) section
$\Delta\phi$: incremental curvature
$\Delta\phi_{C,cr}$: incremental curvature due to unrestrained creep
$\Delta\varepsilon_{cen.cr}$: incremental unrestrained creep strain at the centroid of tr section
$\Delta\varepsilon_{c,sh}$: incremental unrestrained shrinkage strain at the centroid of tr section
$\Delta f_{p,R,k}$: incremental change in stress in each prestress layer k due to relaxation
A_c, A_{tr}: area of concrete and tr section, respectively
$A_{p,k}$, $A_{s,k}$: total strand area in prestress and reinforce steel layers k, respectively
$y_{p.k}$, $y_{s.k}$: distance from prestressing and reinforcing steel layer k to the centroid of tr section, respectively
E_c, E_p: modulus of elasticity of concrete (time-variable) and strands, respectively
n_p, n_s: modular ratio with respect to prestress and reinforcing steel, respectively
I_{tr}: tr section moment of inertia

Then, the incremental strains at different girder layers are obtained with the consideration of perfect bonding between the steel and the concrete. The incremental stress distribution is calculated based on the linear-elastic stress-strain behavior.

In Fig. 2d, the computed incremental strains and stresses are summed up with the total values from the previous time step to obtain the total strains and stresses. In Fig. 2e, camber development at each time is calculated from the curvatures (i.e., by use of the moment-area theorems); strain and stress distributions are computed with curvatures and axial strains. To represent the observed conditions at the precast-prestressed plants, girders are assumed to be symmetrical and simply-supported and to have a uniformly distributed self-weight.

3 Experimental Data

The data for the bottom-flange strains, midspan curvatures, and midspan cambers were gathered from four research projects covering fifty-one full-scale prestressed girders designed and produced in accordance with the US practice. The span length and the hardened concrete properties are provided in Table 1. The girders in Project 1 were fabricated with HPC for a bridge in Tallapoosa County, Alabama [6]. The girders in Project 2 were produced for a research study with different concrete strengths [6, 12]. The girders in Project 3 were fabricated with HPC for a bridge in Macon County, Alabama [1]. The girders in Project 4 were fabricated with a range of span lengths and concrete strengths for research purposes [13].

Camber measurements were completed by using surveying instruments in all of the projects. In Project 1, a tensioned-wire system was also used. The bottom-flange strains were monitored using the vibrating-wire strain gages (VWSG) as well as the temperature histories.

Table 1 Span lengths and hardened material properties

Project No & Info*			Girder#	Span length (m)	$f_{c,i}$ (MPa)	$f_{c,28}$ (MPa)	$E_{c,i}$ (GPa)	$E_{c,28}$ (GPa)
(1)	BT-54, H-S	VC	7	29.6	57.8	70.1	47.1	49.0
		SCC	7	29.6	60.1	72.2	42.9	44.5
(1)	BT-72, H-S	VC	7	40.8	57.6	74.7	47.1	50.1
		SCC	7	40.8	54.5	72.6	40.1	43.5
(2)	Type I, M	VC	2	12.2	33.0	47.6	39.3	48.4
(2)	Type I, M-S	SCC	2	12.2	38.2	67.5	36.2	51.4
(2)	Type I, H-S	SCC	2	12.2	71.9	92.2	48.3	58.3
(3)	BT-54, H-S	VC	5	34.1	58.9	68.4	39.6	39.6
(4)	T-Beam, M	VC	3	4.0, 4.9, 7.0	34.5	43.6	33.8	35.5
(4)	T-Beam, M-A	SCC	3		37.9	58.9	33.8	37.2
(4)	T-Beam, M-S	SCC	3		36.5	63.2	34.1	47.9
(4)	T-Beam, H-S	SCC	3		68.9	92.3	41.7	48.6

*Girder types are standard AASHTO shapes except for the customized T-beams in Project 4. The labels M and S represent moderate- and high-strength concrete, respectively. The labels A and S represent the addition of fly ash and slag cement, respectively. The labels VC and SCC represent vibrated and self-consolidating concrete, respectively

The girders in Projects 1, 2, and 3 were stored outside and affected by a non-uniform temperature distribution over the cross-section. The raw strain and camber readings were adjusted relative to an internal temperature of 20°C for a uniform temperature distribution. The internal strain readings were corrected for the temperature change inside the VWSG and for the nonlinear temperature distribution over the concrete cross section. The corrections for camber measurements were made in two steps: (1) the baseline correction for the initial camber due to the significant temperature gradient reading and (2) correction for the temperature gradient at each camber reading. For interested readers, the details of prestressed steel layout, girder geometry, testing methodology, and corrections were available in the referred documents with a summary in [7].

The midspan readings are averaged for each group and are given in Table 2. The girders in Projects 1 and 2 include the temperature corrections. Curvature readings were determined from the multiple strain readings located farthest from each other over the sectional depth. Initial readings represent the time immediately after the prestress transfer for Projects 1, 2, and 4, and one day after the prestress transfer in Project 3. The difference between the total readings and the initial readings gave the *growth* readings. The chronological ages of girders at the erection vary between 90 to 353 days after prestress transfer. The deformation *growth* is between 50% and 75% of the initial measurements in Project 1 regardless of its type. In Project 2, the *growth* is above 100% of the initial strain measurements and the *growth* of curvature and camber depend on the concrete type. In Project 3, the *growth* is between 20% and 60% of the initial measurements regardless of its type. Few measurements were available in Project 4.

4 Results

To compare the predictions with the observed data, the numerical predictions are performed with the use of actual test data available for each of the fifty-one girders. *fib* MC 2010 is based on the European construction practice, while AASHTO LRFD and ACI 209 are based on the US practice. The equivalent input parameters are needed. Type III cement having high early strength was used in all of the projects. Therefore, "42.5 R, 52.5 N, 52.5 R" is selected as the strength class of cement for *fib* MC 2010. The aggregate types were #67, #7, and #78 limestone, and the type of "basalt, dense limestone aggregates" is selected as the aggregate type for the use of the prediction model of *fib* MC 2010.

For the use of the prediction model of ACI 209, 'steam curing' is selected for the girders with accelerated curing since temperature is elevated above normal curing temperature due to the exposure of sun, tarp/enclosure, or the external application of steam or heat. The AASHTO LRFD shrinkage predictions are performed without the 20% increase.

The girders are then numerically segmented into forty cross sections linearly up to midspan. The time period is divided into forty logarithmic time steps up to the

Table 2 Average of measured bottom-flange strain, curvature, and camber at midspan

Project No & Info*			Strain ($\times 10^{-6}$ m/m)			Curvature ($\times 10^{-4}$ 1/m)			Camber (mm)		
			Initial	*Growth* to		Initial	*Growth* to		Initial	*Growth* to	
				56-day	Erection		56-day	Erection		56-day	Erection
(1)	BT-54, H-S	VC	−439	−239	−279	−2.60	−1.44	−1.57	26	16	17
		SCC	−437	−228	−287	−2.72	−1.42	−1.58	28	13	15
(1)	BT-72, H-S	VC	−463	−231	−338	−1.78	−1.33	−1.28	34	24	22
		SCC	−489	−238	−374	−1.73	−1.16	−1.21	36	25	24
(2)	Type I, M	VC	−256	−349	−408	−4.20	−3.93	−3.93	7	6	7
(2)	Type I, M-S	SCC	−222	−308	−399	−3.14	−2.83	−4.45	6	4	5
(2)	Type I, H-S	SCC	−167	−173	−196	−2.13	−1.20	−2.03	4	1	3
(3)	BT-54, H-S	VC	−879	−228	−519	−6.22	−1.41	−2.36	73	32	34
(4)	T-Beam, M	VC	–	–	–	–	–	–	5	4	4
(4)	T-Beam, M-A	SCC	–	–	–	–	–	–	5	4	4
(4)	T-Beam, M-S	SCC	–	–	–	–	–	–	5	2	–
(4)	T-Beam, H-S	SCC	–	–	–	–	–	–	4	–	–

*The system of labeling is same as Table 1. For the girders in Project 4, some measurements were not performed (shown with "–")

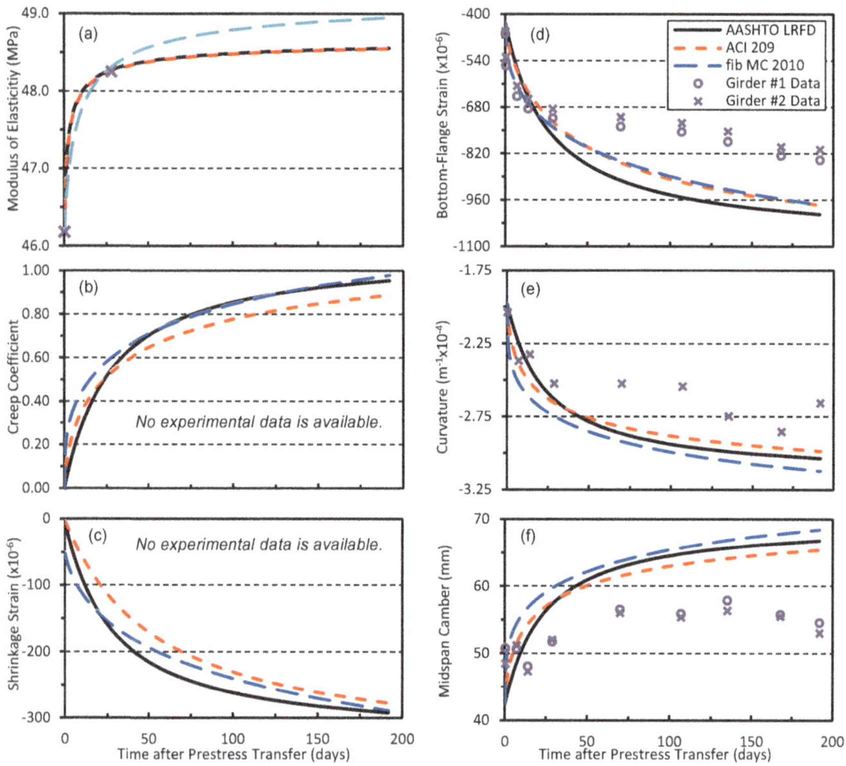

Fig. 3 Measurements versus predictions for the BT-72, H-S VC girder in Project 1

erection age. The bottom-flange strains, midspan curvatures, and midspan cambers are predicted with the concrete MOE development model in Eq. 1.

The output of the ALCAMBER v1.0 software and the observations for the BT-72, H-S VC girder in Project 1 are provided in Fig. 3. The graphs on the left column show the input material model and the graphs on the right column show the resulted time-dependent deformations at the midspan. Negative strain represents the contraction; negative curvature indicates concave-down flexure. Positive camber represents the upward deflection.

In Fig. 3a, the MOE curves obtained by AASHTO LRFD and ACI 209 intersect with the tested MOE at transfer and at 28 days. *fib* MC 2010 has a differed MOE development curve due to the equivalent age (i.e., the effect of the early-age temperature) [11].

In Fig. 3b, the creep deformations are zero at transfer and it begins developing immediately after the prestress transfer. In Fig. 3c, the shrinkage predictions are zero at transfer except for the *fib* MC 2010 shrinkage model since it separately calculates the autogenous and drying shrinkage and uses the equivalent age. Therefore, the shrinkage predictions till the time of first strain measurements (i.e., the benchmark of strain readings) were eliminated for a reasonable comparison.

In Fig. 3d, the bottom-flange strains (at the same depth as the strain gage in the bottom flange) represent the loss of prestress forces since most of the strands are located in the vicinity of the bottom-flange strain gage. In the developed code, the predictions of bottom-flange strain *growth* depend on shrinkage, creep, and the distance to the centroid of the transformed section, y (i.e., $\Delta\varepsilon_{cen} + y\Delta\phi$). The prestress forces are accurately predicted up to 30 days and overpredicted in magnitude at later ages, mostly by AASHTO LRFD.

In Fig. 3e and f, the predictions and measurements of midspan curvatures and camber are plotted. Curvature measurements at midspan are utilized to trace camber predictions (which are predicted by the integration of curvatures along the girder) [7]. It should be noted that the predictions of curvature and camber growth obtained by Eq. 3 depend largely on the creep-induced curvature for the considered girder shapes. Shrinkage deformations and prestress losses have a relatively less impact on the camber *growth*. The initial curvature and camber are accurately predicted up to 30 days and overpredicted (in magnitude) at later ages.

From Figs. 3d, e and f, it can be seen that the use of shrinkage and creep models gives inaccurate time-dependent deformations relative to the field measurements.

The accuracy of the predictions by different creep and shrinkage models is quantified by fractional error, S:

$$S = \sqrt{\frac{1}{n-1}\sum_{i=1}^{n}\Delta_i^2} \tag{4}$$

Where, n: number of data points

Δ_i: fractional error of the ith data point defined as:

$$\Delta_i = \frac{x_{predicted} - x_{measured}}{x_{measured}} \tag{5}$$

Where, $x_{predicted}$: predicted strain, curvature, or camber

x_{measured}: measured strain, curvature, or camber

Table 3 summarizes the fractional errors of AASHTO LRFD, ACI 209, and *fib* MC 2010 on the prediction of bottom-flange strain, midspan curvature, and midspan camber. The closer the fractional error is to zero, the more accurate the prediction is. The errors in *growth* are provided to better discern the accuracies of the creep and shrinkage models since the errors rooted in the initial readings and the predictions are eliminated.

Initial errors, on which the creep and shrinkage models have a negligible effect, are very similar. The initial predictions are predicted more accurately than the *growth* predictions in most of the cases. Initial readings may be affected by the hold-down locations restraining the upward movement immediately after the transfer. The initial errors are mainly increased by the data in Project 3, available at 1 day after the

Table 3 Fractional error of predictions with different creep and shrinkage models

	Strain predictions			Curvature predictions			Camber predictions		
	Initial	*Growth* to		Initial	*Growth* to		Initial	*Growth* to	
		56-day	Erection		56-day	Erection		56-day	Erection
Data Amount	38	38	36	26	26	24	51	48	45
AASHTO LRFD	0.10	0.89	0.68	0.22	0.29	0.18	0.30	0.64	0.54
ACI 209	0.09	0.85	0.82	0.22	0.30	0.21	0.30	0.61	0.51
fib MC 2010	0.09	0.71	0.74	0.22	0.23	0.21	0.30	0.68	0.64

transfer. If the specified design concrete properties (which are typically less than the observations) were used as in the current practice, the initial errors could be more significant as also demonstrated in [8].

For AASHTO LRFD and ACI 209, the *growth* predictions to 56-day show less accuracy than the *growth* predictions to erection. For *fib* MC 2010, the *growth* errors to 56-day and to erection are within the similar range. The *growth* predictions of curvature are more accurate than the *growth* predictions of camber and strain. At erection, the predictions of curvature and strain *growth* to erection are most accurately estimated by AASHTO LRFD, and the predictions of camber *growth* are most accurately made by ACI 209.

Overall, the initial prestress losses (based on the bottom-flange strains) are obtained accurately with an error of about 9%. The *growth* of prestress losses to erection is predicted with an error more than 68%, which is frequently overpredicted as reported in [7] for the same data. The initial camber predictions are obtained with 30% error. Camber *growth* to erection is predicted with an error more than 51%. Both initial and *growth* cambers are mostly overpredicted as reported in [7].

A number of limitations could have influenced the results. The specimens used for the material testing may not be representative of whole girder properties, i.e., the imperfections. Also, the accuracy of *fib* MC 2010 may be affected by the assumptions on its coefficients.

5 Conclusions

The AASHTO LRFD, ACI 209, and *fib* MC 2010 creep and shrinkage models do not accurately predict the *growth* of prestress losses and the camber *growth* despite the use of actual material properties and complex incremental time-steps method. The most accurate predictions for the *growth* of prestress losses are made by ACI 209 with an error of 51% at erection. The camber *growth* is predicted most accurately by AASHTO LRFD with an error of 68%.

The regional calibrations, which accounts for the regional variability in the manufactured process, are not considered in this study. The regionally adjusted models (e.g., [8]) could improve the accuracy of the *growth* predictions.

Acknowledgements Funding for the research by the Alabama Department of Transportation is gratefully appreciated. The authors are thankful to Kurtis M. Boehm, Jon Browne, Brandon R. Johnson, Sam Eskildsen, Kelly R. Levy, Dr. Anton K Schindler, Claire E. Schrantz, and Dr. J. Michael Stallings for their inputs to the experimental data or ALCAMBER v1.0. The contents of this paper reflect the views of the authors who are responsible for the facts and accuracy of the data presented.

References

1. Stallings JM, Barnes RW, Eskildsen S (2003) Camber and prestress losses in alabama HPC bridge girders. PCI J 48(5):90–104
2. ACI 435 (1995) Control of Deflection in Concrete Structures (ACI 435R, Reapproved 2000). Farmington Hills, Michigan: American Concrete Institute, 435R-21 – R435-25
3. Barr PJ, Angomas F (2010) Differences between calculated and measured long-term deflections in a prestressed concrete girder bridge. J Perform Constructed Facil 24:603–609
4. Tadros MK, Fawzy F, Hanna KE (2011) Precast, prestressed girder camber variability. PCI J 56(1):135–154
5. Schrantz CE (2012) Development of a User-Guided Program for Predicting Time-Dependent Deformations in Prestressed Bridge Girders. MS Thesis, Auburn University
6. Johnson BR (2012) Time-Dependent Deformations in Precast, Prestressed Bridge Girders. MS Thesis, Auburn University
7. Isbiliroglu L (2014) Predicting Time-Dependent Deformations in Prestressed Concrete Girders. MS Thesis, Auburn University
8. Mante DM, Barnes RW, Isbiliroglu L, Hofrichter A, Schindler AK (2019) Effective strategies for improving camber predictions in precast, prestressed concrete bridge girders. Transp Res Rec 2673(3):342–354
9. AASHTO (2014) AASHTO LRFD Bridge Design Specifications. 7th ed. Washington, DC: American Association of State Highway and Transportation Officials (AASHTO)
10. ACI 209 (1992) Prediction of Creep, Shrinkage, and Temperature Effects in Concrete Structures (ACI 209R-92). (Reapproved 1997). Farmington Hills, MI
11. *fib* (2010) Model Code for Concrete Structures 2010. Lausanne, Switzerland: International Federation for Structural Concrete
12. Boehm KM, Barnes RW, Schindler AK (2010) Performance of Self-Consolidating Concrete in Prestressed Girders. Highway Research Center Research Report: Auburn University
13. Levy, KR (2007) Bond Behavior of Prestressed Reinforcement in Beams Constructed with Self-Consolidating Concrete. MS Thesis, Auburn University

Bridge Engineering Optimization Opportunities Through Integrated Solutions: Design – Constructive Method

Pedro Pacheco, Pedro Borges, Hugo Coelho, and Diogo Carvalho

Abstract The demand for productivity and cost optimization is a common requirement of bridge construction process nowadays. Focusing on span-by-span concrete bridges, different solutions have proved to be efficient. The choice for the most adequate solution is not linear and depends on several factors of quite different nature, such as bridge geometry (curvature, transversal slope, longitudinal slope), deck section (weight, shape), span length and bridge total length, environmental conditions, site logistics, local construction and design traditions and constructor experience and preferences. Independently of the chosen solution, a fundamental factor for achieving a cost and time-efficient construction process is a timely cooperation between the bridge designer and the construction process developer, making key adaptations in bridge design in order to make it easier for construction. In this specific case, details do matter and a judicious choice may have a significant impact in construction times. In this paper, different span-by-span concrete bridge deck construction processes are presented, ranging from segmental pre-cast to cast in situ solutions. The different solutions have a common denominator – the demand for productivity. The presentation is based on real examples. Throughout the text, a reference is made to Organic Prestressing System (OPS), an actively controlled prestressing system that increases structural efficiency. In its recent applications on bridge construction equipment, OPS confirmed a positive impact in productivity. A particular focus is given to the M1-90-S, a span record holder movable scaffolding

P. Pacheco (✉)
Department of Civil Engineering, FEUP, Porto, Portugal
e-mail: pedro.pacheco@berd.eu

P. Pacheco · H. Coelho · D. Carvalho
Engineering & Production Department, BERD, Matosinhos, Portugal
e-mail: hugo.coelho@berd.eu

D. Carvalho
e-mail: diogo.carvalho@berd.eu

P. Borges
Technical Solutions Department, BERD, Matosinhos, Portugal
e-mail: pedro.borges@berd.eu

P. Gülkan et al. (eds.), *Developments in International Bridge Engineering*, Springer Tracts on Transportation and Traffic 17, https://doi.org/10.1007/978-3-030-59169-4_7

system, mainly manufactured in Turkey and currently in operation in Ankara-Sivas high speed railway.

1 Introduction

The demand for productivity and cost optimization is a common requirement of bridge construction process nowadays. Focusing on span by span concrete bridges, different solutions have proved to be efficient. The choice for the most adequate solution is not linear and depends on several factors of quite different nature, such as bridge geometry (curvature, transversal slope, longitudinal slope), deck section (weight, shape), span length and bridge total length, environmental conditions, site logistics, local construction and design traditions and constructor experience and preferences.

Independently of the chosen solution, a fundamental factor for achieving a cost and time-efficient construction process is a timely cooperation between the bridge designer and the construction process developer, making key adaptions in bridge design in order to make it easier for construction. In this specific case, details do matter and a judicious choice may have a significant impact in construction times.

In this paper, different span by span concrete bridge deck construction processes are presented, ranging from segmental pre-cast to cast in situ solutions. The different solutions have a common denominator – the demand for productivity.

One important aspect regarding this design-constructive Method approach is that recent technological developments enabled, in the last years, new Limits in Medium-large multi Span Viaducts.

This problem or challenge to the worldwide bridge engineering community is not new and a plethora of studies and papers can be found discussing different approaches on this matter. Important studies are presented [1–3] but several other examples could be given.

2 What Key Issues for the Constructive Method Selection

It is known that main factors of decision, while selecting a Constructive Method are: Economy; Time and Safety, not necessarily by this order. Economy is normally evaluated by the sum of Material Consumption; Main Equipment; Man Power; Secondary Equipment; Site Preparation; Indirect Costs and Others.

Each construction method has its own merits and, most probably, each method has a preferred field of application. Surely the "best" decision should be taken in a case by case basis. The adoption of a construction method is part of a "triangular decision" of conceptual design, which always comprises Structural System (& Cross Section) – Material – Constructive Method. Several well-known factors and several well-known conditions are to be considered in each case. Among all factors, there

Table 1 Qualitative evaluation of common constructive methods in multi span viaducts.

Constructive method	Common span limits	Technical validation (>70 m)	Cost	Time	Durability
Prestressed Concrete (PC) - Ground Scaffolding	40–50 m	B-C	C	C	A
PC Precast Beams	40–50 m	C	A-B	A	A-B-C
PC Span by Span Precast Segments	50 m	A-B*	A-B	A	A-B
PC Incremental launching	70 m	B-C*	B-C	A-B	A
PC Precast full segments (box girders)	70 m	A-B *	B-C	A	A
PC Cast in Situ, Span by Span (MSS)	70 m	A*	A	A-B	A
Continuous Composite or Steel Girders/Trusses (diverse methods)	>100 m	A	B-C	A-B	A-B-C
PC Cantilever Method - Precast segments (Launching Gantry)	>100	A	B-C	B	A-B
PC Cantilever Method - Precast segments – lifting	>100	A	B-C	B-C	A-B
PC Cantilever Method – In situ	>100	A	B-C	C	A

Legend: A-Good; B–Acceptable; C–Critical; *-with identified recent progress

are four that have a systematic importance: Technical Validation, Cost, Time and Durability.

Let us do an exercise of qualitative evaluation of the most common constructive methods used for multi span large viaducts (see Table 1):

The observation of Table 1 allows to stablish a main conclusion: each bridge is to be carefully studied as each bridge deserves one solution. Each construction method has its own merits and certainly there are bridges where such method fits better that other methods. But the opposite is also truth (Fig. 1).

Then, in a first stage one should verify the Technical Validation of each method for each bridge potential construction method. Only the other fundamental issues are to be evaluated. It is relevant to enhance here that in the last 20 years relevant technological achievements occurred, making possible to use methods that were previously "not validated" for larger spans and in particular, for spans above 70 m.

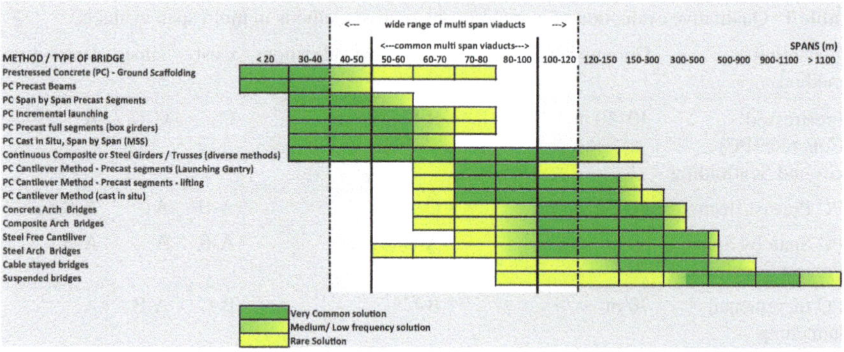

Fig. 1 Span ranges for the most common constructive methods/bridge types in the beginning of the XXI century [4]

3 New Limits in Medium-Large Multi Span Viaducts

Moreover, in what concerns to the particular case of multi-span large viaducts, where there is a strong need for high industrialized solutions, there is clearly a "critical range of spans" – 70 to 100 m. Indeed, within that span range, only one prestressed concrete (PC) industrialized method was commonly applied (PC Cantilever Method – precast segments - with launching gantry). Below the limit of 70 m, several methods were common or even very frequent.

Most probably these 3 methods, presented in Fig. 2, are still among the most adequate in the present state of art for multi-span large viaducts (even some, with several decades). Nevertheless, the third is typically more used for a higher range, than the mentioned "critical range of spans", and the second is typically used more for a span bellow that range.

In few words, there is a "critical range of spans" – 70 to 100 m – where the adequacy of the construction methods seems restrictive. Meaning that great part of multi-span viaducts comprise spans equal or above 70 m. This means that in several cases the span definition may be conditioned by the construction method. In other words, most probably, in several cases, the optimal span may be "refused" not by design criteria, but due to construction method restrictions or inadequacy.

Very recently some of these methods had relevant developments and become now effective options for all players in bridge engineering international community [4].

Certainly, there are now less restrictions for decision makers, and most probably, in several cases, this span range increase will contribute for the adoption of the most rational and competitive solution, when such method is the most adequate.

Mainly, this represents another degree of freedom for bridge designers, for constructors and for project owners (Fig. 3).

Fig. 2 Examples of construction methods in Multi-Span Large Viaducts, from the top to bottom: Ponte Rio Niteroi, Brazil – PC cantilever method – precast segments, launching gantry, 80 m span; Ponte Vasco da Gama, Portugal – PC precast full segments, 60 m span; Benicia-Martinez Bridge, USA – PC cantilever method – precast segments, lifting. Spans: 127–201 m

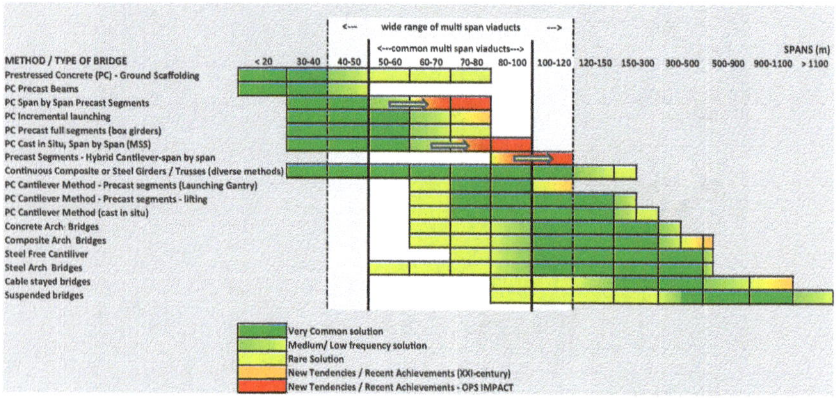

Fig. 3 Actual span ranges for the most common constructive methods/bridge types [4]

4 The Opportunity of Material Saving Due to Construction Method Change/Adoption

Frequently the current span definition of a multi span bridge or viaduct is a result of previous experiences or internal market limitations in construction equipment availability. Indeed, "similarity" may be one of the most common criteria to define the span length of a particular viaduct. But in several cases, this approach may be far of the optimal criteria, in what concerns material optimization.

One of the typical exercises of material consumption optimization is the study the "optimal span" of a bridge or viaduct. That exercise consists in a set of preliminary calculation on materials consumption for Foundations + Piers & Deck, for different spans. In Fig. 4 the essentials of an "optimal span" study carried out for the Ana Garibaldi Bridge are presented (second example of optimization in Table 5).

In what regards Material Consumption, one of the main factors is the Prestressing Consumption. It is known that for the same deck cross section and for the same spans, the prestressing consumption may have relevant variations.

Recent studies were developed regarding the material consumption of different methods [5–7]. One of them [5] presents a comparative analysis of prestress consumption of a 75 m multi-span concrete box-girder road bridge constructed with different methods: Movable Scaffolding System (with joint at the support and at a distance of the support of 1/5 of the length of the span), large precast segments (with or without span continuity) and small precast segments. In some of the construction methods the deck is considered simply supported.

In that study, in order to obtain a more targeted comparison, the bridge geometry is the same regardless the construction process.

For each construction method, the prestress parameterization is based on existing projects and studies. In each case, the quantity of prestressed cables obtained is the minimal that verifies all conditions required. The mentioned study, conducted according to the Eurocodes, consists in a static elastic analysis and considers the viscoelastic behavior of the materials.

The materials' time-dependent behavior and the influence of the construction phases and redistribution of bridge stresses and forces are taken into account. The stresses are analyzed during the construction and service stages of the bridge. The service phase of the bridge is analyzed at the end of the construction and 10 000 days after.

The final output of that work is the prestressing quantities at a general span for each construction method and a comparison between those quantities. No other economic factors are taken into account, neither the materialization of the prestressing cables (Fig. 5).

This study is no more than a reference and obviously is developed for a very restrict span range, more preciously, just for a specific span. But surely, it enables

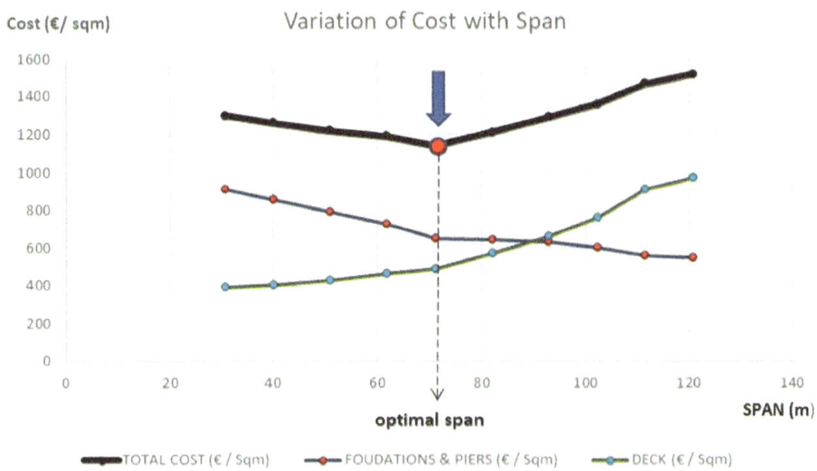

Fig. 4 Study on "optimal span" considering materials consumptions in different elements for Ana Garibaldi Bridge

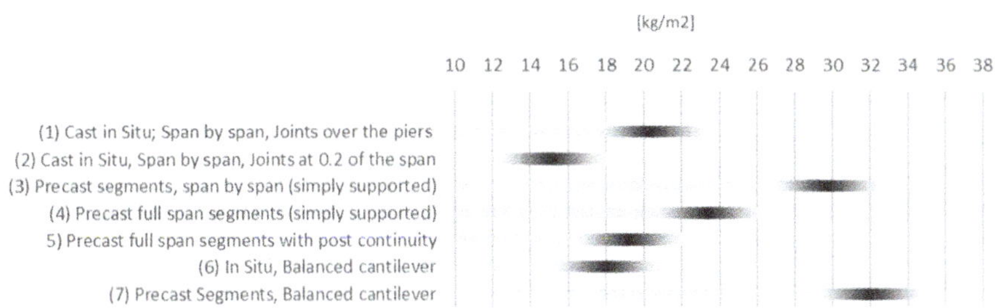

Fig. 5 Study on comparation on prestressing steel consumptions in 7 different constructive methods for multi-span viaducts of 75 m [5]

to feed qualitative studies as per Table 1, and more important, enables bridge engineers to understand the relevant material consumptions differences among different construction methods.

Indeed, if one compares the fundamental diagrams of flexural moments of different methods, as in Table 2, it becomes obvious that very significant material consumptions are to occur.

Obviously, the related typical prestress layouts are to be coherent and significantly different for each construction method type. In Table RRR the typical prestressing layouts for different constructive methods are presented [5] (Fig. 6).

The authors have different experiences worldwide with active participation in viaducts and bridges designed for different constructive methods (precast segmental - span by span; cast in situ – span by span; cantilever method – in situ; ground scaffolding; precast girders; etc.). In 2011, it was identified that one of the methods that was not properly chosen, several times, was the construction with movable scaffolding system – span by span – cast in situ. A specific study on this method was then carried out and published [8]. Some of the essentials of this study are as follows.

The construction of bridge and viaduct decks with several spans with movable scaffolding systems (MSS) may be a very efficient and competitive constructive method. This solution is generally used for the 40–60 m span range. Over the last few years' new experiences have been made and new solutions have been developed for the range 70–90 m (large MSS or LMSS). In this range surprising economical results may be achieved if the number of spans is high and/or if the costs of piers and foundations are relatively high. With LMSS it is possible to achieve very high productivity ratios.

Regarding these technological steps, an important role is to be given to the OPS (Organic Prestressing System) Technology, which after the initial steps was implemented in the beginning of this century [9, 10] with success in the following years [8, 11–13]. This was an important contribution to ensure the reliability and effectiveness of the application of Movable Scaffolding Systems (span by span in situ construction) and Launching Gantries to limits not possible before.

These technologies proved to be extremely reliable [14] and with very relevant results in terms of operational results for constructors [8] and with relevant positive impact in sustainability [15]. Other similar applications were also proved to be effective [12, 13] (Table 3).

These studies and sequent studies published later [4] were fundamental to make possible the first worldwide application of Movable Scaffolding Systems for 90 m multi-spans. This application happened in Turkey in 4 viaducts included in the High Speed Railway Line between Ankara and Sivas and the equipment M1-90-S, designed by BERD, performed the world record of span dimension for in situ multi-span construction.

Further than time reduction of deck construction cycle – from about 90–100 days/ form-traveller pair/ span to 14 days/span, a very significant materials consumption optimization was provided. Indeed, the change of constructive method – from cantilever method to continuous span by span deck – enables very relevant concrete and steel savings in the deck with very high influence on the materials optimization

Table 2 Constructive methods and structural systems characterization [5]

Method / Structural System Evolution – Lateral view	Main Equipment
(1) Cast in Situ; Span by span, Joints over the piers 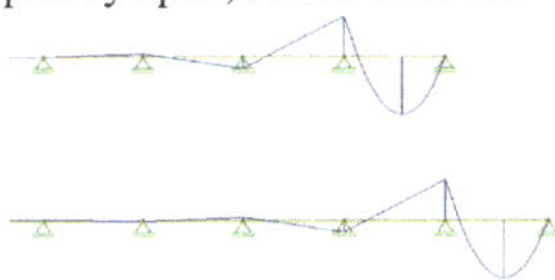	 Figure 1A. MSS (BERD, 2010)
(2) Cast in Situ, Span by span, Joints at 0.2 of the span 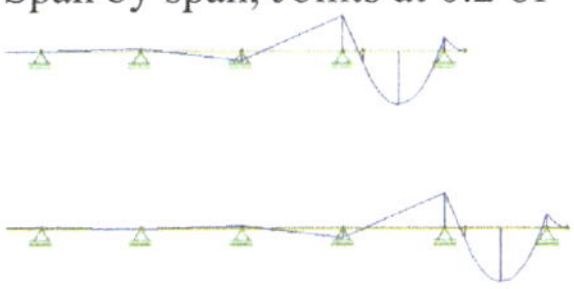	
(3) Precast segments, span by span (simp. Supported 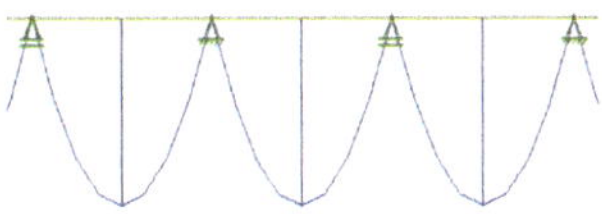	 Figure 1B. Launching Gantry (BERD, 2014)
(4) Precast full span segments (simply supported) 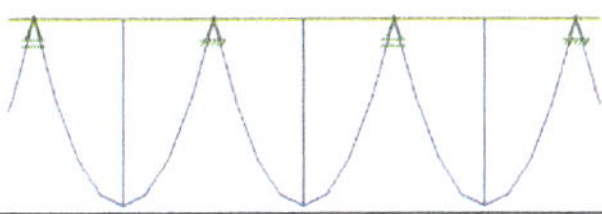	 Figure 1C. Mega Crane - Central spans of Vasco da Gama Bridge, Lisbon, (Vinci, 1997)
(5) Precast full span segments with post continuity 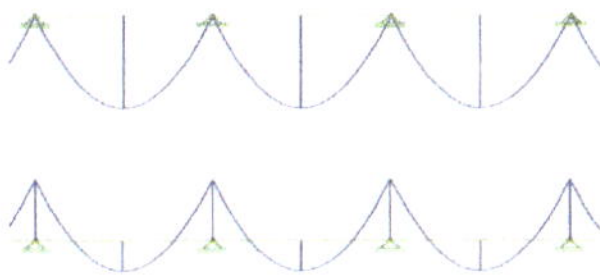	
(6) *In Situ*, Balanced cantilever 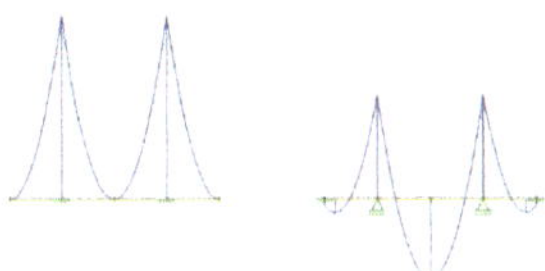	 Figure 1D. Cantabria, Spain (Construgomes 2007)
(7) *Precast Segments,* Balanced cantilever 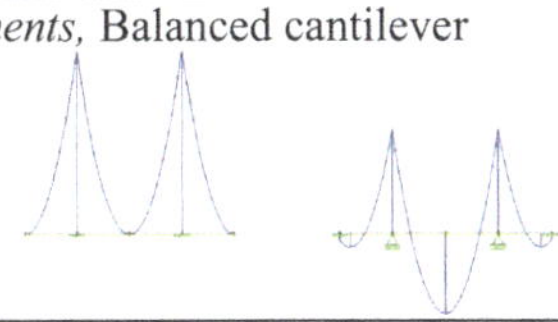	 Figure 1E. Tenerife, Spain (Ferrovial, 2012)

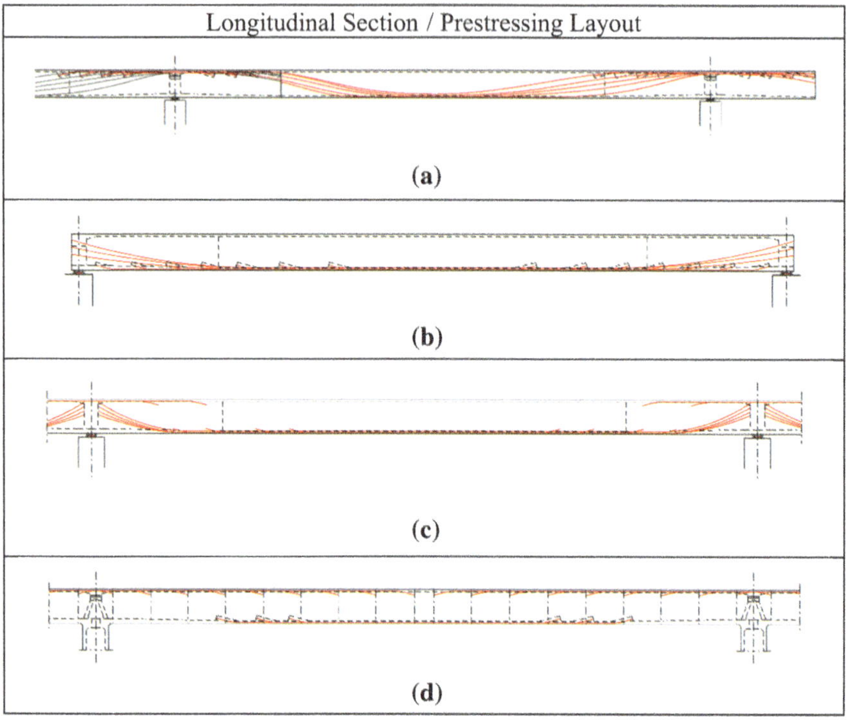

Fig. 6 **a** Prestressing layout at method (2) Span by Span, in Situ, Joint at 0.2 L **b** Prestressing layout at methods (3) and (4) Simply Supported **c** Prestressing layout at method (5) Precast full span segments with post continuity **d** Prestressing layout at methods (6) and (7) Balanced cantilever

Table 3 Bridge design recommendations – for span by span, in situ multi -span viaducts conceptual design [8]

1	Maximum actual span with LMSS - 90 m
2	Location of joints – at L/4
3	Prestressing layout – classical span by span solution
4	Consideration of Horizontal loads at front piers: (1) horizontal projections of LMSS weigh; (2) LMSS sliding friction; (3) Braking loads; (4) forces in locomotion reaction points; (5) Wind actions on LMSS and (6) accidental LMSS induced forces
5	Consideration of adequate combinations of Horizontal Forces and Vertical forces, according to LMSS functioning
6	Eventual consideration of wind induced vibrations on piers, eventual need of LMSS bracing
7	Consideration of accidental LMSS induced actions on the deck

Table 4 Materials consumption optimization – Ankara-Sivas HSR LINE – Viaducts V7, V9, V10 and V15.

Structural Elements	Original Solution - (cantilever method)		Adopted Solution - (span by span – in situ)		OPTIMIZATION
Foundations					
Concrete (piles & blocks)	332976	m3	146546	m3	56%
piles	34714	m3	27915	m3	20%
blocks	298262	m3	118631	m3	60%
Steel	48250	ton	25417	ton	47%
piles	8284	ton	5018	ton	39%
blocks	39966	ton	20398,8	ton	49%
Piers					
Concrete	126476	m3	67857	m3	46%
Steel	35116	ton	17637	ton	50%
Deck					
Concrete	117300	m3	81474	m3	31%
Steel	33076	ton	25202	ton	24%
Prestressing steel	4617	ton	4288	ton	7%
GLOBAL WEIGHTED OPTIMIZATION					39,55%

in the piers and foundations due to the importance of mass to the seismic action assessment. Main results on the optimization are presented in Table 4 – based on information of the tender and from site.

These numbers are extremely relevant and reveal an enormous innovation capacity of all players, including Designers, Sub-Contractor, Contractor and the Turkish Railway authority in HSR – TCDD. This is a remarkable contribution for the bridge engineering state-of art, worldwide, but it is also an enormous contribution for sustainability.

Considering the unitary costs adopted for this project, the partial optimizations in all structural components lead to a Global Weighted Optimization of 39.55% in the 4 viaducts of this project.

5 Some Worldwide Experiences in Optimization Solutions Adoption

In following Table 5 some experiences in optimization solutions, worldwide, are presented.

Table 5 Experiences of integrated solutions optimization: design – constructive method

	Country / Original Method / Adopted Method / Type of Optimization
	**Country** – Slovakia **Original Method** - In Situ, Double deck/ 42 m span **Adopted Method** - In Situ, Unique deck/ 69 m Span **Type of Optimization:** Materials on Piers / Time Reduction
	**Country** – Brazil **Original Method** - Precast segmental, cantilever Method **Adopted Method** - Precast segmental, Span by span **Type of Optimization:** Time Reduction: 50 m/ 2 weeks to 100 m/week
	**Country** – Brazil **Original Method** - Precast Beams **Adopted Method** - In situ, Span by span **Type of Optimization:** Time Reduction: 25 m/week to 60 m /week
	**Country** – Mexico **Original Method** - Incremental Launching **Adopted Method** - In situ, Span by span **Type of Optimization:** Prestressing consumption: less 40% (estimated value)
	**Country** – Turkey **Original Method** - In situ, Cantilever **Adopted Method** - In situ, Span by span **Type of Optimization: Time** - one span cantilever method 90-100 days per form traveller set; with MSS lasts 14 days. **Construction materials savings** - average reduction of 39.55%

6 Conclusion

Independently of the chosen solution, a fundamental factor for achieving a cost and time-efficient construction process is a timely cooperation between the bridge designer and the construction process developer, making key adaptions in bridge design in order to make it easier for construction. In this specific case, details do matter and a judicious choice may have a significant impact in construction times or in material consumptions.

In this paper, different span by span concrete bridge deck construction processes are presented, ranging from segmental pre-cast to cast in situ solutions. The different solutions have 3 common denominators – the demand for productivity, economic efficiency and sustainability. In the 5 presented Projects relevant and different goals are achieved regarding that denominator. Probably the last, implemented in Turkey, is the most paradigmatic example, as it brings simultaneously relevant advantages in terms of time, economic indicators and sustainability.

Acknowledgements The authors wish to deeply thank the Designers, Sub-Contractor and Contractor who participated in the present project and in particular KMG, KAPPA, DOGUS and with high importance TCDD who decided to go ahead and contribute for an upgrade in the worldwide state of art, in the technological, economic and sustainability perspectives. Finally, the authors would like to thank the specialized consultants that also had important contributions for this project, namely the company YUKSEL PROJE, Prof. Alp Çaner and Prof. Murat Dicleli.

References

1. Mathivat J (1980) The cantilever construction of prestressed concrete bridges, 1st Spanish edition, EDT, S.A.
2. Majewski L (1976) Bauing, DasVorschubgerust fur die Ahrtalbrucke Der Bauingineur, vol 51, pp 25–28. Springer-Verlag
3. Matt P (1983) Status of segmental bridge construction in Europe. PCI J 28:104–125
4. Pacheco P (2015) Multi span large decks – the organic prestressing impact. In: Multi-span large bridges, pp 103–124. CRC
5. Lopes F, Ferreira A, Lima B et al (2015) Comparative study of presstressing consumptions in 7 different constructive methods for 75 m multi-span box girders. In: Multi-span large bridges, pp 507–513. CRC Press
6. Figueira de Sousa F (2018) Comparação de Consumos de Pré-esforço em Tabuleiros Pré-esforçados de Betão Armado – Soluções in situ versus soluções com aduelas pré-fabricadas FEUP. Master thesis
7. Morim M (2008) Estudo do tabuleiro de um viaduto de betão pré-esforçado construído tramo a tramo com vãos de 90 m FEUP. Master thesis
8. Pacheco P, Coelho H, Borges P et al (2011) Technical challenges of large movable scaffolding systems. Struct Eng Int (J IABSE) 21(4):450–455
9. Pacheco P, Adão da Fonseca A (2001) Organic prestressing. J Struct Eng 400–405
10. Pacheco Organic prestressing – and example of and effector system. Struct Concr (J Fib) (3):107–113
11. Pacheco P, Guerra A, Borges P et al (2007) A scaffolding system strengthened with organic prestressing – the first of a new generation of structures Structural Engineering International. J IABSE 17(4):314–321
12. Pacheco P, Coelho H, Resende A (2014) Achieveing high productivity in bridge construction – the organic prestressing impact. In: 9th international conference on short and medium span bridges (Calgary, Canada)
13. Pacheco P, Resende A, Torres A et al (2018) Span by span segmental construction: OPS impact in the LG50-S Laguna case. In: 40th IABSE symposium:S17-9-16

14. Pacheco P, André A, Borges P et al (2010) Automation robustness of scaffolding systems strengthened with organic prestressing. Autom Constr 19(1):1–10
15. Pacheco P, Adão da Fonseca A, Resende A et al (2010) Sustainability in bridge construction processes. Clean Technol Environ Policy 12(1):75–82

Load Distribution and In-Plane Superstructure Movements on Highly Skewed Steel Girder Bridges

Mauricio Diaz Arancibia and Pinar Okumus

Abstract Highly skewed girder bridges experience modified load paths, deck diagonal and acute corner cracking, and superstructure horizontal movements caused by long-term loading. This paper presents the instrumentation and load testing of a 23-year-old, three-span, medium-span-length, steel girder bridge with a skew angle of 47° to understand the effects of existing deck cracks on bending and shear girder load distribution. Measured and predicted load distribution factors as per AASHTO LRFD Bridge Design Specifications were compared. Load testing data, consisting of bending and shear strains at multiple locations across the length and width of an exterior span, were used to validate finite element models. Bridge models under temperature loading were used to evaluate the importance of bridge skew angle on in-plane superstructure displacements. The investigation showed that skew had a major role in bridge in-plane displacements, and consequent superstructure in-plane rotation, leading to greater transverse displacements with increasing skews under thermal loading.

1 Introduction

Bridges are complex structures that are critical to a nation's transportation system. They are designed to cross over natural or man-made obstacles such as waterways, roadways, and railways. Skewed bridges are needed when these obstacles run oblique to the centreline of the feature carried by the bridge, resulting in a rhombus-shaped bridge superstructure in plan. Skew is defined as the smallest angle between a line perpendicular to the bridge centreline and bridge support lines (Fig. 1).

Despite the need for high skew bridges, it is documented that skew bridges experience several construction and service problems that may worsen with increasing skew angles [1, 2]. Bridge engineers have been aware of the detrimental effects of high skew angles on bridge performance since the early 1900's as documented by Waddell [1] in his book "Bridge Engineering" in 1916. The thirteenth item in Waddell's list

M. D. Arancibia · P. Okumus (✉)
University at Buffalo, The State University of New York, Buffalo, USA
e-mail: pinaroku@buffalo.edu

P. Gülkan et al. (eds.), *Developments in International Bridge Engineering*, Springer Tracts on Transportation and Traffic 17, https://doi.org/10.1007/978-3-030-59169-4_8

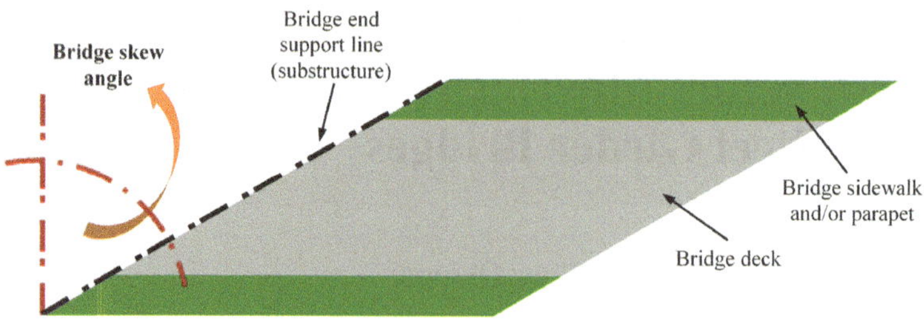

Fig. 1 Skew angle on the plan view of a bridge

of "First Principles of Designing" states that "The building of a skew-bridge should always be avoided when it is practicable". Research and field observations over the years on skewed bridges have proven Waddell right. High skew changes load paths for live loads [3–8]. Bridges with high skew are also reported to have increased cracking in the deck and substructure, large in-plane superstructure displacements, girder bearing distress, and superstructure misalignment with the substructure.

Although AASHTO LRFD Bridge Design Specifications [9] have provisions to adjust live load distribution between deck and girders, it does not provide guidance on performance issues observed with high skew bridges over long term. In studies on long term effects (i.e., shrinkage and temperature) on bridges, high skew was not the focus or studies on bridge displacements on high skewed bridges have not covered bridge superstructure in-plane rotations [10–12].

This paper discusses load distribution of steel girder deck bridges through the load testing of a high skewed steel girder bridge, and finite element analyses validated by the test data. The finite element analyses were used to investigate the long-term behaviour and service problems commonly associated with high skew bridges.

2 Load Testing of a High Skew Steel Girder Bridge

The bridge selected for load testing was a 23-year-old, three-span, continuous, concrete deck-on-steel W-girders bridge in Chippewa Falls, Wisconsin, USA. It had a skew angle of 47°. The span lengths were 14 –21 m 14 m (47–70–47 ft). The deck width was 13 m (42 ft). The girders were 84 cm (33 in.) deep. Girder bearings were elastomeric at the expansion joint, and metallic at the joint restraint against translation. The straight superstructure had a concrete, 20 cm (8.0-in.) thick cast-in-place deck, composite with girders except over the piers. The concrete deck rested on six steel girders with a constant spacing of 2.2 m (7.25 ft). A photograph of the bridge showing the large skew is given in Fig. 2 on the left.

Girders are connected transversely through channel-shaped steel diaphragms along spans, over piers and abutments. The specified concrete compressive strength for the deck and substructure is 28 MPa (4 ksi) and 24 MPa (3.5 ksi), respectively.

Fig. 2 The steel girder bridge with a high skew used for load testing (left) and one of the girders instrumented with electrical resistance strain gages (right)

Girders and diaphragms were ASTM A572 Grade 50 and A36 steel, respectively. Mild steel reinforcement has a specified yield strength of 414 MPa (60 ksi).

The bridge carried highway traffic and had existing cracks in the deck providing an opportunity to understand the impact of deck cracks on load distribution. Test results were also used to validate finite element models that were later employed for a parametric study.

Short-term loading consisted of two fully loaded dump trucks with 5 axles moving from one end of the bridge to another at crawl speed (i.e., less than 5 mph) to simulate static loading on the bridge. Trucks were placed on the bridge in eight load configurations. These configurations were created to place the highest loads on exterior and interior girders, in one-lane and multiple-lane loading scenarios. Each load configuration was repeated twice to ensure that the bridge response readings were reliable. During each pass, trucks stopped at approximately the mid-span of the instrumented span to achieve a fully static loading at a known position.

Instrumentation on the bridge included strain gages that measured shear and bending on all girders along several locations along one span of the bridge. Load test results were used to understand live load distribution between girders, and to validate the finite element models. Figure 2 on the right shows the resistance strain gages installed on one girder on the bottom flange and on the web.

Figure 3, on the left, shows the influence diagrams for bending strains measured at three different locations across the bridge span under the loading created by two trucks in tandem. The truck location across the width of the bridge is also shown on the figure. In this figure, lines with different colours belong to strain measurements from different girders.

Figure 3, on the right, shows the moment live load distribution factors (DF) calculated based on the measured bending strains. This figure also includes the live load distribution factors calculated using the AASHTO LRFD Bridge Design Specifications. Influence line and live load distribution results for other load cases are provided elsewhere [13].

The results showed that the three girders nearest to loading carried more than 75% of the load. For all load cases moment and shear distribution from the deck to girders

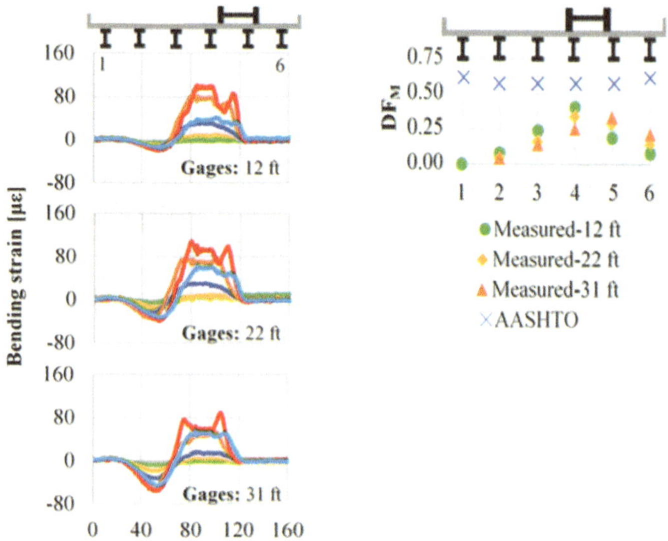

Fig. 3 Bending strain influence diagrams for each girder (left), live load moment distribution factors (DF_M) calculated from the measured strains (right). 1 ft = 0.30 m

was safely predicted using the AASHTO LRFD Bridge Design Specifications. For cases where wheel loading was applied directly over the girder, moment predictions from AASHTO LRFD Bridge Design Specifications did not have a large safety margin.

3 Finite Element Analyses

3.1 Analysis Features, Material Properties and Loading

The bridge used for load testing was analysed using the finite element method. Both 1-dimensional (1-D, girder-line) and 2-dimensinal (2-D) models of the bridge were created. The 1-D models are used commonly in bridge design practice due to their simplicity. They are composed of a single frame element under HL-93 loading. They are used with the live load distribution factors given by the AASHTO LRFD Bridge Design Specifications.

2-D models were created to evaluate the accuracy of the 1-D models, since skew can change load paths and live load distribution. In addition, 2-D models were created to investigate the impact of long term loading on bridge performance. 2-D models were composed of frame elements to model girders, and shell elements to model the deck. The deck and the shell elements were connected to each other with body constraints. The deck and shell elements were offset from each other to capture the

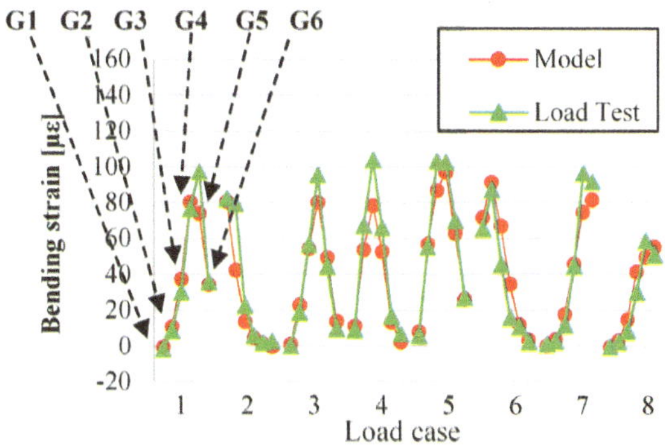

Fig. 4 Comparison of bending strains obtained from models and testing

correct positions of each bridge component. Girder bearings were modelled using link elements. Secondary bridge elements such as intermediate diaphragms were not included in the models.

Mechanical properties of concrete deck and steel girders were input as given by the AASHTO LRFD Bridge Design Specifications. Loading composed of HL-93 loading to capture the effects of live loads. To understand the impact of skew on long term behaviour of the bridge, temperature loading was applied to the bridge superstructure. The temperature loading was taken from the Wisconsin State Bridge Manual [14], which provided thermal changes recommended for bearing and expansion joint designs. Thermal changes were ±24 °C (75 °F) for steel bridges.

3.2 *Validation of Analysis Results*

In order to validate the finite element model, bending and shear strains obtained through analysis were compared to the strains measured through load testing. For validation, the loading used in the finite element models replicated the one used for the load test. Figure 4 shows a comparison of the bending strains 3.7 m (12 ft) away from the bridge end on each girder (G1-6) as obtained from the load testing and finite element modelling for all load cases.

Figure 4 shows that there was an acceptable correlation between the finite element models and the test results. Results of the finite element models for bending and shear strains on other locations along the span were similarly comparable to the test results, are available elsewhere [13] and are not shown here for brevity.

4 Analysis Results on Live Load Distribution

The bridge tested had a skew angle of 47°. Finite element models of this bridge was modified to understand the effect of skew on live load distribution. Skew angle was

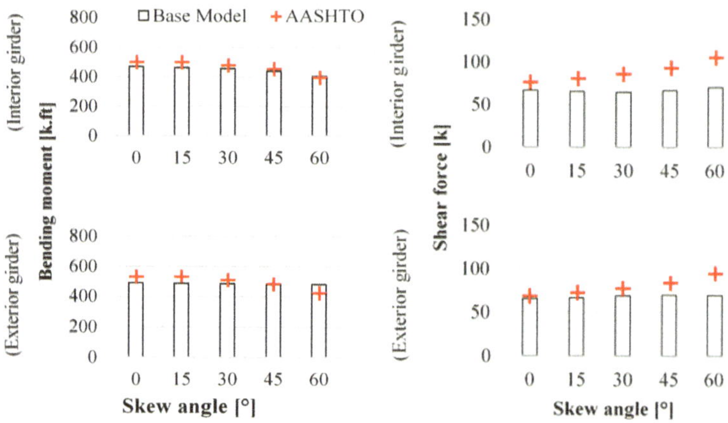

Fig. 5 Moment and shear reactions in girders with varying skew

varied between 0° and 60° (the limit of girder-line analyses allowed by AASHTO) in 15° increments. Bending moment and shear reactions obtained for varying skew angles are given in Fig. 5. The same figure also compares the reactions obtained using the AASHTO distribution factors. The results show that girder moment reactions decrease and shear reactions increase with increasing skew. AASHTO predictions of the bending moment on exterior girders was slightly un-conservative for the 60° skew bridge. All other predictions were conservative.

5 Analysis Results on the Long Term Response

Finite element analyses were performed for the bridge used for load testing with varying skew angles to understand the impact of skew on bridge superstructure in-plane displacements and rotation. Loading considered was a seasonal negative temperature change of 24 °C (−75 °F).

Figure 6, on top, displays transverse bearing displacements at each bridge corner for the steel bridge tested, for varying skew angles under negative temperature loading as obtained from finite element analyses. The figure reveals a nonlinear relationship between transverse displacements and bridge skew angle. Displacements at the obtuse corners were consistently smaller than the ones at the acute corners. Transverse bearing displacements significantly increased with increasing skew angles relative to the ones of the non-skewed counterpart. At 60° skew, transverse displacements at acute corners were more than 10 times of the ones for the non-skewed bridge.

The displacements shown in Fig. 6, on bottom, also indicate that bridges with high skew tend to rotate in plane under temperature loading. For a negative change in temperature, bridge rotation is towards obtuse corners as shown in the figure on the bottom. The rotation is expected to be clockwise for positive temperature changes.

Fig. 6 Transverse displacements at the girder bearings with varying skew

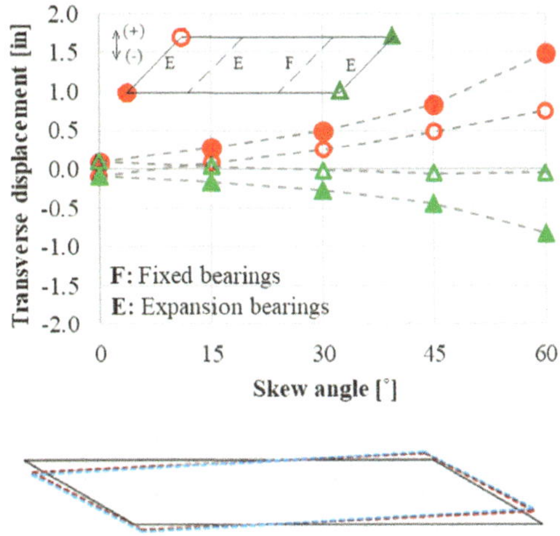

This rotation leads to the misalignment between superstructure and substructure components, as documented by field inspections.

6 Summary and Conclusions

This study investigated the short and long-term effects of skew on steel girder deck bridges through load testing and finite element modeling validated by load testing. The following are the main conclusions of this study:

- The live load distribution factors provided by the AASHTO LRFD Bridge Design Specifications and 1-D (girder line) analyses can predict load distribution of highly skewed deck bridges with a reasonable margin of safety. This margin of safety was the smallest for moment as obtained from load testing. Finite element models of bridges with varying skew angles showed that AASHTO predictions can be slightly un-conservative for moment for bridges with 60° skew.
- Simple 2-D analyses that do not include secondary bridge members were able to predict the live load response of steel bridges. These linear-elastic analyses are also computationally efficient, and are recommended to compliment girder line analyses for high skew bridges.
- Under temperature loading (long-term), transverse displacements of high skew bridges were much larger than the ones with no or small (<20 °) skew. Transverse bearing displacements at the acute corners were larger than the ones at the obtuse corners of high skewed bridges. This causes bridges to rotate in the superstructure plane in the counterclockwise direction under negative temperatures and pushes superstructures out of alignment with the substructure. Girder bearings and expansion joints should be designed to accommodate these displacements.

Acknowledgements This research was funded through the Wisconsin Highway Research Program by the Wisconsin Department of Transportation and the Federal Highway Administration under Project 0092-16-05. The contents of this paper reflect the views of the authors who are responsible for the facts and accuracy of the data presented herein. The contents do not necessarily reflect the official views of the Wisconsin Department of Transportation or the Federal Highway Administration at the time of publication.

References

1. Diaz Arancibia M, Okumus P, Oliva MG (2017) Review of skew effects on prestressed concrete girder bridges: problems and current practices. In: Conference proceeding, PCI convention and national bridge conference, Cleveland, OH, 28 February–4 March, Paper, vol 50, p 18
2. Diaz-Arancibia M, Okumus P (2018) Load Testing of Highly Skewed Concrete Bridges. ACI Special Publication on Evaluation of Concrete Bridge Behavior Through Load Testing - International Perspectives, SP323–2, 18
3. Waddell JAL (1916) Bridge engineering. Wiley, New York
4. Nutt RV, Schamber RA, Zokaie T (1988) Distribution of wheel loads on highway bridges. National Cooperative Highway Research Program (NCHRP), Project report 12–26, Washington, DC
5. Huang H, Shenton WH, Chajes JM (2004) Load distribution for a highly skewed bridge: testing and analysis. J Bridge Eng 9(6):558–562
6. Barr PJ, Amin MN (2006) Shear live-load distribution factors for I-girder bridges. J Bridge Eng 11(2):197–204
7. Bou Diab F, Mabsout M, Tarhini K (2011) Influence of skew angle on live load moments in steel girder bridges. J Bridge Struct 7(4):151–163
8. Nouri G, Ahmadi Z (2012) Influence of skew angle on continuous composite girder bridge. J Bridge Eng 17(4):617–623
9. American Association of State Highway and Transportation Officials (AASHTO) (2014) AASHTO LRFD bridge design specifications, 17th edn, Washington, DC
10. Roeder CW, Moorty S (1991) Thermal movements in bridges. Transp Res Rec 1290:135–143
11. Moorty S, Roeder CW (1992) Temperature-dependent bridge movements. J Struct Eng 118(4):1090–1105
12. Tindal TT, Yoo CH (2003) Thermal effects on skewed steel highway bridges and bearing orientation. J Bridge Eng 8(2):57–65
13. Diaz Arancibia (2018) Design and performance of highly skewed deck girder bridges. PhD dissertation, University at Buffalo, the State University of New York, 337
14. WisDOT Bureau of Structures (2015) Highway Structures Information (HSI). https://on.dot.wi.gov/dtid_bos/extranet/structures/LRFD/index.htm. Accessed Dec 2014

Soil-Structure Interaction Analysis of a Railway Bridge in Rotterdam

Cenan Ozkaya

Abstract Theemswegtrace bridge is a double track railway bridge located in Rotterdam, Netherlands. Total length of the bridge is 3.7 km, with a typical span length of 30 m. For these typical spans, superstructure consists of simply supported precast prestressed I girders and cast in place R/C slab. An assessment study of the original design was performed in a non-linear manner to observe the importance of soft soil conditions on rail-structure interaction. Due to soft-soil conditions, rail-structure interaction analyses according to UIC 774-3 were performed together with inclusion of barrette piles and associated soil-springs. Site specific non-linear p-y curves, t-z curves and q-z curves were developed and used in the analyses. It was found out that soils structure interaction can effect the results significantly and has remarkable importance in rail structure interaction analysis of bridges located on soft soils.

1 Introduction

The viaduct has a total length of 3.7 km, with a typical span length of 30 m. There are double tracks with normal speed railway traffic. Width of the platform is E = 11.5 m. No rail expansion device for typical spans was selected in the original design. Within the scope of the work, rail-structure-soil interaction analyses for a bridge segment having a total length of 485 m were performed. Some steel-arc bridge segments are out of scope for this study. Bearing system for the studied segment consists of mobile elastomeric bearings and mobile elastomeric bearing + dowel, providing horizontal fixity for each simply supported segment. The client was Ballast Nedam, Netherlands.

Results for this segment shall be indicative for the rest of the viaduct. Typical cross-sectional view of the superstructure and sub-structure are presented in Figs. 1 and 2.

C. Ozkaya (✉)
Yüksel Proje International, Ankara, Turkey

P. Gülkan et al. (eds.), *Developments in International Bridge Engineering*, Springer Tracts on Transportation and Traffic 17, https://doi.org/10.1007/978-3-030-59169-4_9

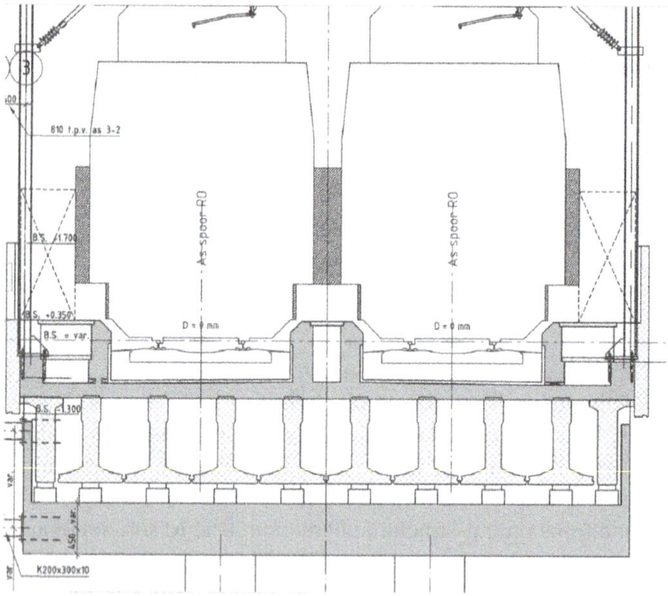

Fig. 1 View of the superstructure cross-section

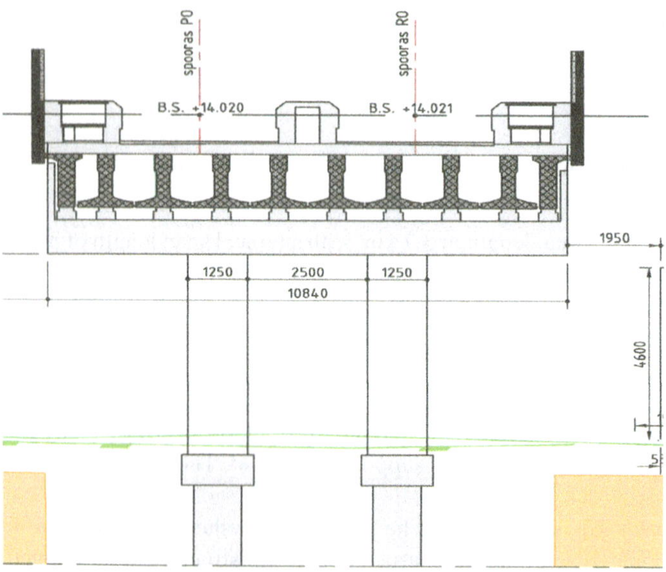

Fig. 2 View of the superstructure and sub-structure

For the superstructure, the system consists of precast girders and the girder depths change from 1.22 to 2.2 m. Piers have rectangular shape with dimensions 1.25 × 2.5 m, except 3–4 axes with 1.5 m × 3 m pier dimensions. Within the segment studied, 3 types of barrettes and 2 types of caissons are used. Barrettes types have dimensions of 1 × 7.5 m, 1 x 5 m, 1 × 1 0 m and caissons have 5 × 3 m and 3.75 × 3 m dimensions with 1 m wall thickness.

In the studies, Eurocode 1991-2 [1], UIC 774-3 [2] relevant National Annexes and international codes as well as documents provided by the client [3] are taken as basis. The loads and load combinations as well as other relevant parameters are taken from the above documents to be able to fully check the original design. Larsa 4D [4] software is used for structural analysis.

2 Analysis Model and Results

2.1 Substructure with Barrette Piles

In the model, superstructure, rails, piers, barettes and caissons are modeled by using frame elements. Bearings are modeled by using spring elements. Dowels providing fixity is modeled by introducing very high horizontal stiffnesses at those locations although behaviour of dowels is questionable, especially in long term.

The barettes and caissons are included in the model since their flexibility resulted from the loose soil conditions is the dominant factor affecting the rail-structure interaction response of the viaduct. Geotechnical studies performed for this project at Middle East Technical University are taken as basis for non-linear p-y, t-z and q-z spring load–displacement curves. Soil layers at the site consist of loose to medium stiff sand. There is no bedrock at the bottom.

In the analysis model these soil springs as well as springs representing the behaviour of ballasted track are modelled by nonlinear bi-linear springs. A general view of the analysis model is presented in Fig. 3.

Since no rail expansion device is foreseen in the original document for this segment, rails are considered to be CWR (Continuous Welded Rail) and requirements of EN 1991-2 and UIC 774-3 for CWR are followed. Live load classification factor α is accepted as 1.21.

For braking, acceleration and live load analysis, loaded spring properties are used for the ballast whereas unloaded spring properties are used for modelling thermal behaviour of the track.

Absolute horizontal displacement of the deck under braking + acceleration is about 12 mm which exceeds the upper code limit of 5 mm (EN 1991-2 6.5.4.5.2) significantly. Approximately half of this displacement result from the flexibility of the barrettes&caissons. Relative displacements between the superstructure modules, the basis for evaluation, sometimes exceeds the code specified limit of 5 mm. On the other hand, rail stresses are almost within the code limits. Typical rail axial forces under temperature rise of 35° is presented in Fig. 4.

Settlements at the pile bottoms were checked to observe extent of the settlement. The maximum settlement at the tip of the caissons/barettes under service load combination is 15 mm, which is within the acceptable limit of 25 mm. Therefore, no settlement problem is expected. The axial load distribution in barrettes/caissons under representative service load combination is presented in Fig. 5.

As seen from the figure, most of the axial load is resisted by skin friction. Contribution of tip resistance to axial load resistance is quite small, as expected. Large perimeter of the barette piles resulted in high total vertical skin friction resistance.

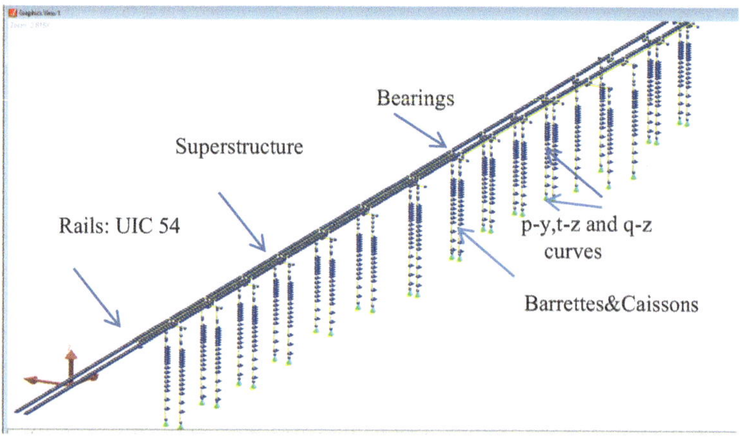

Fig. 3 General view of the analysis model

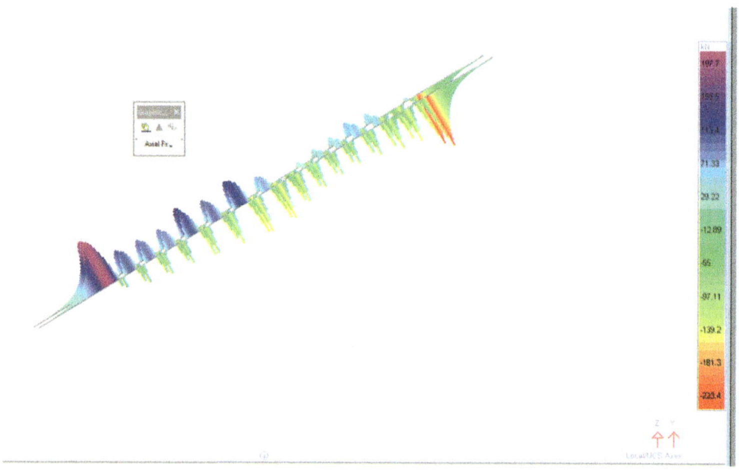

Fig. 4 Rail forces under a temperature rise of 35°

Fig. 5 Axial load distribution in barrettes/caissons under service load combination

Fig. 6 Batter pile alternative with 9 piles

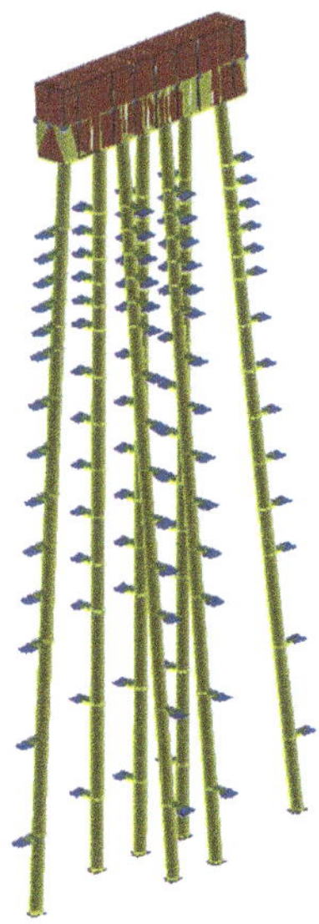

P-y curves behave elastically in their horizontal load displacement curve under braking and acceleration forces.

Table 1 Summary of pushover analysis results

System	Horizontal load (kN)	Horizontal pile cap displacement (mm)
No Batter- 7 Tubex Piles	1000	6.17
7 Tubex Piles	1000	5.59
Two Supplementary 5:1 Inclined Batter Piles-9 piles	1000	4.20

2.2 Batter Pile Alternative

Tubex type batter pile [5] alternative is studied upon the request of the client in detail by performing incremental pushover analyses. The diameter of the piles is 559 mm with a composite cross-section. The thickness of the outer steel shell is 10 mm, with a concrete fill. The general view of the batter pile model is presented in Fig. 6. Soil springs are modeled by using the same p-y curves with consideration of pile dimensions and nodal spacings in the analysis model. There are 7 piles in a pile cap. An alternative with 9 Tubex piles is also studied. The inclination of batter piles is 5:1. The number of battered piles is 2 in 7 pile alternative whereas it is 4 in 9 pile alternative. No batter Tubex pile alternative is also studied.

Analysis is performed by applying incremental horizontal loads at the pile cap level. The horizontal displacement is recorded at the same level.

Batter pile alternative with 7 piles provide almost same horizontal stiffness of the original barette/caisson alternative. The horizontal stiffness provided by 9 pile alternative is higher than all alternatives, as can be observed in Table 1. Therefore, this alternative may be used at locations where higher substructure stiffness is needed and substructure utilities (pipelines etc.) allow.

3 Conclusions

Rail-structure-soil interaction analyses performed on the selected segment segment resulted in horizontal absolute displacements in the order of 12 mm, which is significantly higher than the 5 mm code limit. Since this displacement limit in the codes implies relative displacement between superstructure modules or the displacement between superstructure and abutment, solution of this exceedance may result in addition of supplementary rail expansion devices at bridge abutment transitions and at intermediate positions where there is sharp transitions in substructure stiffnesses resulting from pier heights, pile dimensions, soil conditions etc.

If this alternative is not feasible for a reason, then substructure stiffness can be increased at some certain locations by increasing pier/barette/caisson dimensions. Alternatively, ground improvement methods can be used in between some barettes/pile caps to provide supplementary horizontal stiffness.

The batter pile is a feasible alternative. The horizontal stiffness provided by proposed batter pile system can be significantly increased by adding two more batter piles with 5:1 inclination in the longitudinal bridge direction or by increasing the batter angles of the two 5:1 inclined batter piles in 7 pile alternative. Soil-structure interaction has an important role in rail-structure-interaction analysis of bridges located on soft soils or having foundation systems having low overall horizontal stiffness.

Acknowledgements The author expresses his sincere gratitude to Assoc. Prof. Dr. Mustafa Tolga Yilmaz from METU-Ankara and Mr. Suleyman Eren Dursun for their valuable contributions.

References

1. Eurocode (2003) EN 1991-2 Actions On Structures-Part 2: Traffic Loads on Bridges , English Version
2. UIC 774-3 (2001) Track Bridge Interaction-Recommendation for Calculations, 2nd edn
3. Theemswegtrace Bridge Original Design Documents (2017)
4. Larsa Inc. (2017) Larsa 4D
5. Tubex Piles (2017). https://www.foundationpiledriving.com/tubex-piles/

Reliability Based Safety Level Evaluation of Cable Members for Cable-Stayed Bridges in Turkey

Alp Caner, Nurdan Apaydın, Yeşim Esat, and Burak Kurtman

Abstract Turkey, ranking second on investing in roads after Slovakia among the OECD countries, has recently started to design and construct long-span cable-supported bridges. In recent years, a live load model accurately representing the actual live load uncertainties prevailing in Turkey has been developed using probabilistic methods. The live load model, KGM 45, is now used in the design of short to medium span bridges in Turkey. In a previous nationally funded research, the superstructure of cable-stayed bridges has also been studied with KGM 45 live load model. The focus of this paper on extending the findings of the previous research into new cable-supported long-span bridge live load designs of Turkey. In this concept, both superstructure and cable designs have been studied by a committee formed by the Turkish Highways, KGM. In this paper, we present only the results of cable designs.

1 Introduction

Most of the international bridge design specifications have live load models covering the design of short to medium span bridges with some limited information on long-span bridges. Bridge design for live load needs to consider the crowd vehicle modeling for long-span bridges. Live load models are either developed using a deterministic or probabilistic method. Until now, there is no consensus among international engineers on live load modeling for long-span bridges. Owner's engineers typically select live load model based on their wisdom or comprehensive technical studies.

A. Caner (✉)
Middle East Technical University, Ankara, Turkey
e-mail: acaner@metu.edu.tr

N. Apaydın
Turkish Highways, KGM, Ankara, Turkey

Y. Esat
KGM, Ankara, Turkey

B. Kurtman
Yüksel Proje, Ankara, Turkey

P. Gülkan et al. (eds.), *Developments in International Bridge Engineering*, Springer Tracts on Transportation and Traffic 17, https://doi.org/10.1007/978-3-030-59169-4_10

A probabilistic approach that covers uncertainties of the design and the construction techniques have been adapted by many countries to define live load models. The measure used to calibrate bridge design using the probabilistic method is called reliability index. In defining loads and resistance factors, reliability analysis has been conducted to set reliability index to a constant value to have a uniform possibility of failure for every single bridge independent of its type as shown in Table 1.

The KGM 45 truck model has been used to determine load and resistance factors using a calibration method. In this concept, precast prestressed girders, steel girders, post-tensioned concrete box girder and cable-stayed bridges have been calibrated using the following procedure [4–8].

- Selecting bridge geometry to be analyzed
- Gathering the statistical database for load and resistance parameters
- Forming the load and resistance models
- Choosing the reliability analysis procedure
- Deciding on the target reliability index
- Computing load and resistance factors corresponding to target reliability index

Argınhan [4] and Çelik [5] have studied live load models for precast prestressed girder bridges with span lengths ranging from 25 to 40 m using uncertainties of Turkey. The probabilistic based method indicated that the KGM-45 truck model is representative of Turkish traffic with the reliability of more than 4.3 for these type of bridges. The KGM-45 truck has been developed during a nationally funded research for the Turkish Highways, and live load pattern is given in Fig. 1. The multiple presence factor for KGM 45 is 0.65 for more than four lanes of traffic.

Table 1 Reliability indexes

Specification or guideline	Reliability index	Probability of failure (Normal Distribution)
Eurocode [1] – RC3 (EN 1990)	4.30 (50 years of life) 4.10 (100 years of life)	0.85/100000 2.06/100000
Eurocode [1] – RC2 (EN 1990)	3.80 (50 years of life)	7.23/100000
Eurocode [1] – RC1 (EN 1990)	3.30 (50 years of life)	48.25/100000
AASTHO-LRFD [2]	3.75 (75 years of life)	8.81/100000
TUBITAK-110G093 [3]	4.20 (75 years of life)	1.33/100000

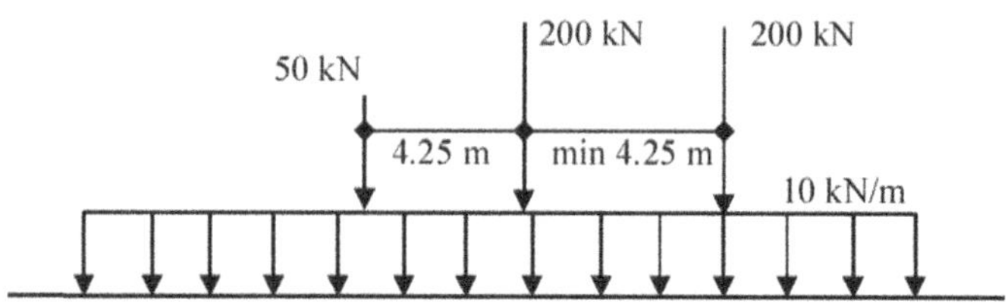

Fig. 1 KGM-45 truck model per lane

Koc [6] has studied steel girder bridges, Çakır [7] has studied post-tensioned concrete box bridges, and Donmez [8] has studied cable-stayed bridges in a very similar way. All these researchers have concluded that KGM 45 truck model both exceed requirements of AASHTO design specifications for the type of bridge of interest.

Reliability index for models with KGM 45 is determined to be above 4.3 that also exceeds Eurocode requirements. In Table 2, a summary of bridges and live load models are given.

The complex traffic congestion can sometimes be over simplified and can be quantified into an equivalent uniform lane load as shown in Fig. 2. For existing bridge designs, variations in live load can be significant. No consensus on the value of live load can be reached as shown in Table 3.

Eurocode [3] designs are more liberal than AASHTO-LRFD [4] designs. The Eurocode live load model shall never be used to check against AASHTO-LRFD design equations or vice versa. The two specifications shall never be mixed up to have a hybrid design. Details of Eurocode LM1 live load model can be found in Celik's

Table 2 Researched bridges with Turkish uncertainties

Bridge type	Researcher	Main span range	Truck models
Precast prestressed girders	Argınhan [4]	25 m–40 m	H30S24, HL-93
Precast prestressed girders	Celik [5]	25 m–40 m	KGM 45, HL-93 LM1
Steel girder bridge	Koc [6]	50 m–80 m	KGM 45, HL-93
Post-tensioned concrete box girder	Cakır [7]	90 m–180 m	KGM 45, H30S24 HL-93
Cable stayed	Donmez [8]	420 m–550 m	KGM 45, H30S24 HL-93

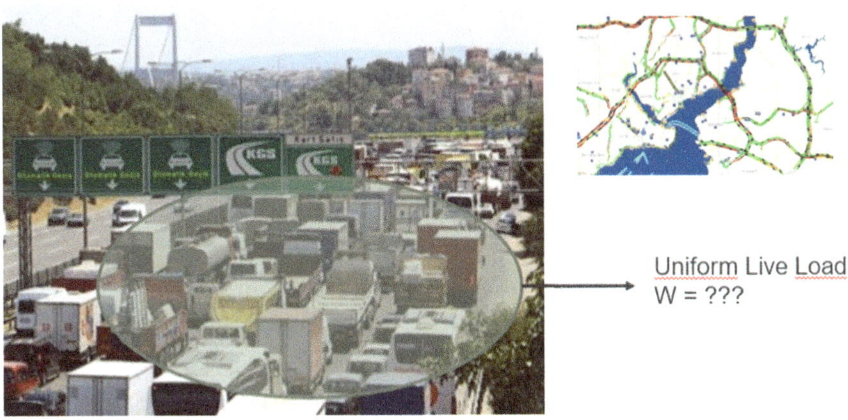

Fig. 2 Complex traffic congestion

research [2]. It shall also be reported that Eurocode live models are developed for bridges with less than 200 m main span.

A comparison between the average of existing designs (Table 2) and design live models are given in Fig. 3. In case, number of lanes are more than four, and main span length is 1000 m, KGM 45 truck model is similar to the USA and Canadian live load model. The Canadian and USA live load models are computed using work of Lutomirska [9].

This study aims to determine the proper reliability indexes for existing designs of cable supported bridges in Turkey. In this concept, only designs of cables under live load combinations are considered.

Table 3 Equivalent average uniform lane load

Bridge	Country	Construction year	Width (m)	Main span (m)	Equivalent uniform lane load (kN/m)
Izmit bay	Turkey	2013	36	1550	11.4
George Washington	USA	1927	36	1098	6.3
Verrazano Narrows	USA	1959	31	1296	5.9
Stonecutters	Hong Kong	2004	50	1015	8.7
Fatih Sultan Mehmet	Turkey	1986	39	1092	6.6
Carquinez Strait	USA	2003	26	724	5.3
Storebealt	Denmark	1995	31	991	2.5

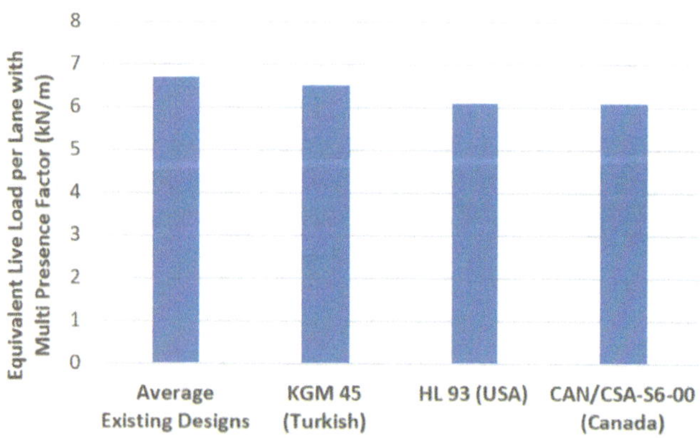

Fig. 3 Equivalent live load for a suspension bridge with main span length = 1000 m and number of lanes > 4

2 Reliability Analysis

Computation of reliability index can be performed using different methods. Method selection is mainly based on target precision, available input data, and computational costs. These methods are:

- Mean-Value-First-Order-Second-Moment (MVFOSM)
- First-Order-Second-Moment (FOSM)
- First Order Reliability Method (FORM)
- Second-Order-Reliability Method (SORM)

The MVFOSM is in the form of an approximate performance function linearized at the mean values of the random variables using a first-order Taylor series, and only utilizes second-moment statistics of the random variables. If failure function is highly non-linear, significant errors may be observed by neglecting higher order terms. Fortunately, in bridge design, such non-linearity is very insignificant, and all methods result in similar values. Therefore, MVFOSM method will be used in this paper for practical reasons.

In this study, mode of failure is selected to be related to cable element load carrying capacity. A traditional notion of the safety limit is associated with the ultimate limit state which is the ultimate axial capacity of the cable.

Let R be the resistance (axial load carrying capacity in this case), and Q represents the load effect (total axial load on cable); then, the limit state function, g, can be expressed by

$$g = R - Q \tag{1}$$

The probability of failure, P_F, is defined using the following equation and shown in Fig. 4

$$P_F = P(R - Q < 0) = P(g < 0) \tag{2}$$

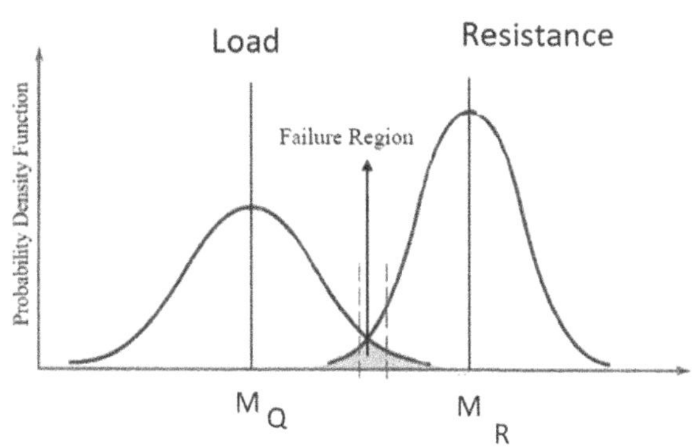

Fig. 4 Probability of failure

The reliability index, β, is determined as a function of P_F

$$\beta = -\Phi^{-1}(P_F) \tag{3}$$

where Φ^{-1} is the inverse standard normal distribution function. The reliability index can be simply computed using the following equation.

$$\beta = (M_R - M_Q)/(\sigma_R^2 + \sigma_Q^2)^{0.5} \tag{4}$$

where M_R is the mean resistance, M_Q is the mean actual load, σ_R is the standard deviation of resistance and σ_Q is the standard deviation of the actual load. The resistance, R, and actual load combination, Q is given in Table 4. Details of truck models are summarized in the study of Donmez [8].

It has also been known that cable break, cable replacement, and seismic cases may sometimes govern the cable design instead of main design load combination having dead and live load. In this study, only the main load combination has been investigated.

The statistical parameters used in this study has been taken from the study of Donmez [8]. A large set of truck data has been analyzed for cable-stayed bridges by Donmez [8] to determine statistical parameters. More than 5000 congestion scenarios have been included. The same statistical parameters have been adopted in this study as shown in Table 5.

Table 4 Resistance and actual load combination

Reference	Truck	Actual load combination, Q	Design resistance, R
AASHTO LRFD	HL-93	Q = 1.00 (DC + DW + LLIM)	$R = 0.65\ A_s\ f_s$
AASHTO LRFD	H30S24	Q = 1.00 (DC + DW + LLIM)	$R = 0.65\ A_s\ f_s$
TUBITAK 110G093	KGM 45	Q = 1.00 (DC + DW + LLIM)	$R = 0.65\ A_s\ f_s$
EUROCODE	BS5400	Q = 1.00 (DC + DW + LLIM)	$R = A_s\ f_s/1.5$
EUROCODE	LM1	Q = 1.00 (DC + DW + LLIM)	$R = A_s\ f_s/1.5$
EUROCODE	TK BRO	Q = 1.00 (DC + DW + LLIM)	$R = A_s\ f_s/1.5$

DC: dead load, DW: superimposed dead load, LLIM: live load plus impact, As: area of cable and fs: the ultimate tensile strength of the cable.

Table 5 Statistical parameters

Component	Coefficient of variation
Resistance of cables	0.03
Dead load	0.10
Live load	0.15

Table 6 Main parameters of bridges

Bridge	Main span (m)	Deck width (m)
İzmir Korfez	270	21.0
Komurhan	380	23.9
Nissibi	400	22.9

3 Case Studies

Three designed cable-stayed bridges have been investigated. Details of these bridges are given in Table 6. The layout and typical cross sections of bridges can be seen in Figs. 5 and 6. In the design of bridges, actual designs of the bridges are used.

The reliability analysis has been conducted for the three bridges with eight different alternatives. The reliability index is determined to be more than min required value of 4.3 as shown in Fig. 7. The results have indicated that the cables of these bridges are designed with some reserve in addition to minimum requirements of the specifications. Izmır Korfez bridge has less safety compared to two other bridges but still significantly safer compared to the minimum requirements.

Usually, additional reserve capacity has been used in Turkey to avoid drastic future maintenance of the bridges as a policy of the KGM. As it has been known, Turkey is in the last position on maintenance of roads among the OECD countries.

The Eurocode design approach typically results in more liberal design compared to the AASHTO-LRFD design approach. Analysis of TK BRO and HL 93 truck models indicate that a bridge with a limit design with these trucks (reliability index = 4.3) will increase the chance of failure under KGM 45 truck loading. It shall be noted that KGM 45 truck is based on uncertainties of traffic in Turkey. Since all the bridges are not designed for minimum limit state, adequate safety is provided for all.

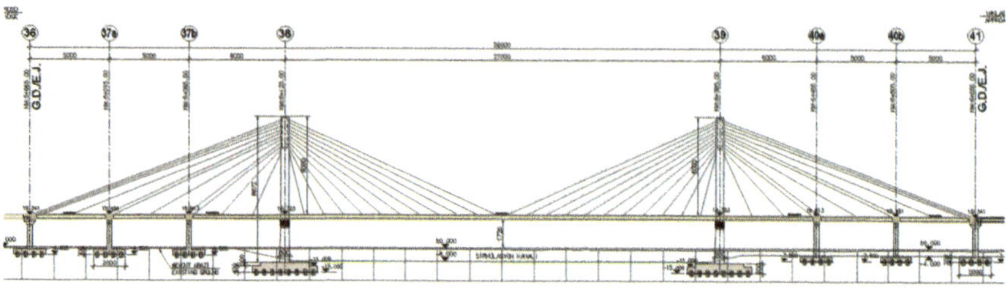

(a) İzmir Korfez

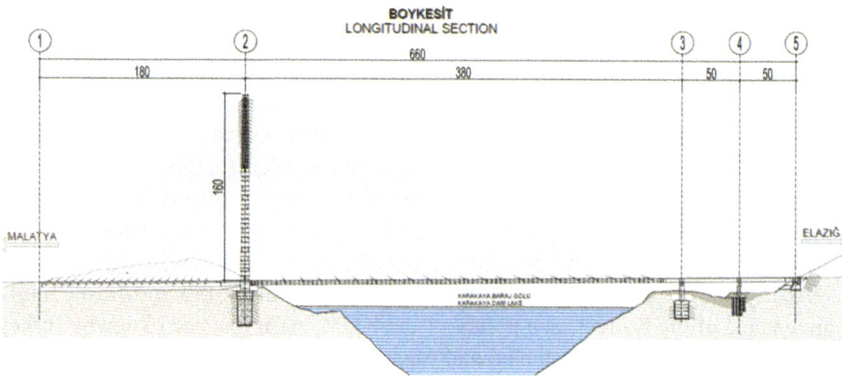

(b) Komurhan

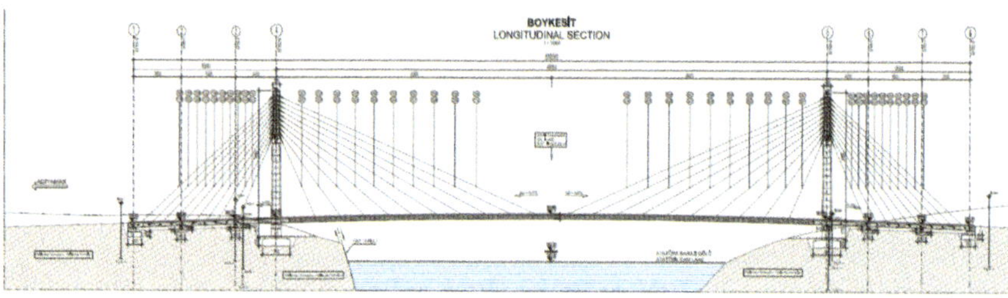

(c) Nissibi

Fig. 5 Longitudinal sections of bridges of interest

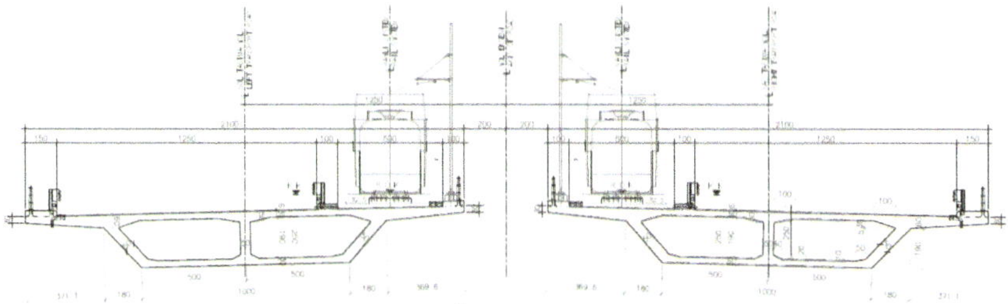

(a) İzmir Korfez

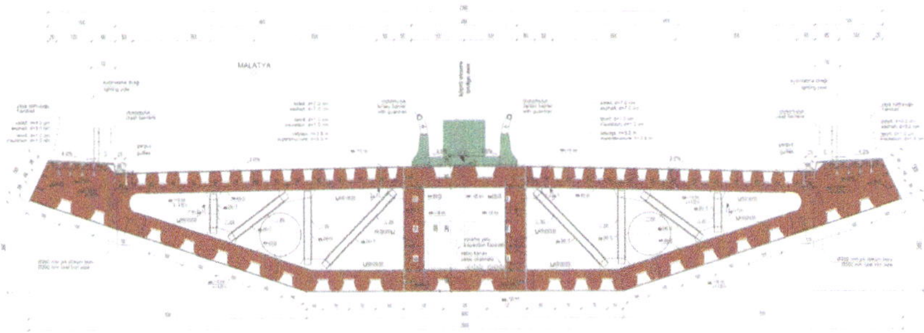

(b) Komurhan

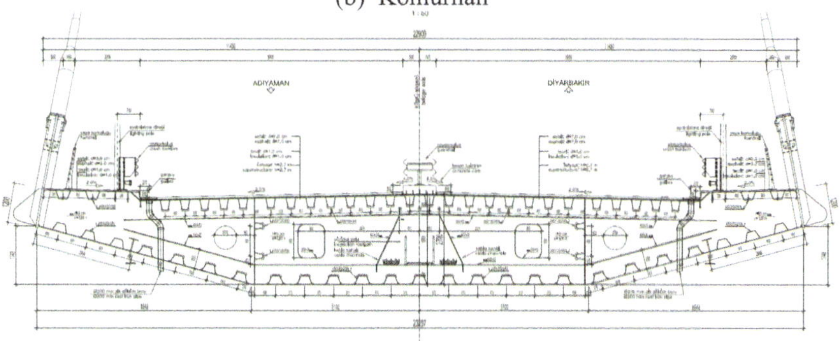

(c) Nissibi

Fig. 6 Typical cross sections of bridges of interest

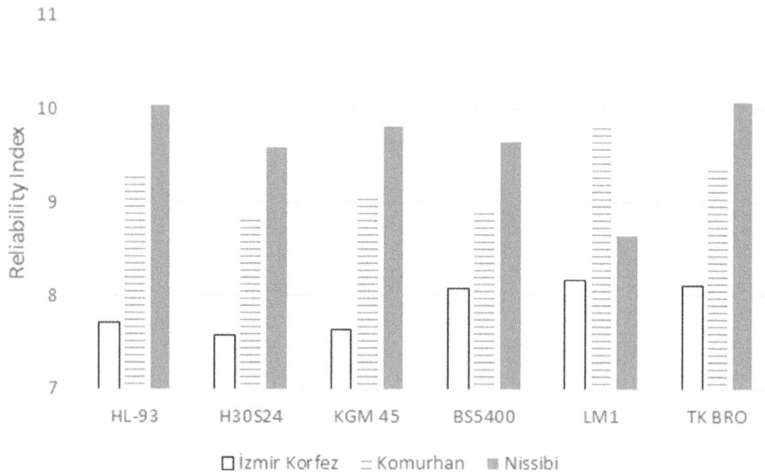

Fig. 7 Summary of results

4 Conclusion

Following conclusions can be drawn from this study.

1. KGM 45 truck model representing uncertainties of traffic in Turkey is appropriate to be used in the design of cable-stayed bridges including cable elements.
2. A limit state design of cable-stayed bridge with an HL-93 truck or a TK BRO truck will have more chance to fail under KGM 45 truck developed for Turkish traffic. It shall also be noted that average existing equivalent uniform load per lane for long-span suspension bridges is closer to the KGM 45 model loads.
3. Cable designs of three bridges have more safety than required. Usually additional reserve capacity decreases future maintenance costs.
4. In general, Eurocode designs are more liberal than the ones performed by AASHTO-LRFD.

Acknowledgements The authors, would like to thank Turkish Highways and Live Load commission for their support.

References

1. AASHTO (2010) LRFD Bridge Design Specifications, 5th Edition, Washington DC, USA
2. Eurocode (2001) EN 1990 Eurocode - Basis of Structural Design, Brussels, Belgium
3. TUBITAK (2014) Development of Bridge Design and Construction Technologies for Turkey, Ankara, Turkey

4. Argınhan O (2010) Reliability Based Safety Level Evaluation of Turkish Type Precast Prestressed Concrete Bridge Girders Designed in Accordance with Load And Resistance Factor Design Method, MS Thesis, Ankara Turkey
5. Celik C (2018) Reliability Analysis of Precast Prestressed Bridge Girders Based On AASHTO LRFD and Eurocode, MS Thesis, Ankara Turkey
6. Koç AF (2013) Calibration of Turkish LRFD Bridge Design Method for Slab On Steel Plate Girders, MS Thesis, Ankara Turkey
7. Cakır BB (2015) Live Load Reliability Index Evaluation for Post Tensioned Balanced Cantilever Bridges MS Thesis, Ankara Turkey
8. Donmez Y (2015) Turkish LRFD Live Load Design Parameters for Cable-Stayed Bridge with Concrete Deck on Steel Girder, MS Thesis, Ankara Turkey
9. Lutomirska M (2009) Live Load Models for Long Span Bridges, Ph.D. Dissertation, University of Nebraska, USA

Design of Izmir Bay Crossing Bridge

Burak Kurtman

Abstract The Izmir Bay Crossing Bridge is a part of a 6.8 km fixed link between Çiğli and İnciraltı districts in Izmir. Total bridge length is 4175 m. Two platforms, each 21 m wide, will combine to carry three lanes of highway traffic and a light railway system. The cable-stayed bridge is located over the circulation channel, and has a total length of 590 m and a main span of 270 m. Pylon height is 88.7 m. The cable-stayed bridge will be constructed by free cantilever method and the rest of the bridge will be constructed by movable scaffolding system. The bridge has a post-tensioned box girder deck with a constant height of 2.5 m. It was particularly important during the design process to ensure structural performance of the bridge because of high seismicity and very poor soil conditions at the site. 2000 mm diameter steel driven piles are used for the foundations. First 20 m of the piles are assumed to be laterally unsupported by soil and pile cap stiffness and mass were considered in the global model for seismic analysis.

1 Introduction

Izmir Bay Crossing project was planned and designed under the tender of General Directorate of Highways. The final (tender) design was completed and approved at the end of 2017. The purpose of the project is to reduce the city traffic, to complete the ring road which is currently not forming a circle around the city and connecting the north and south coasts of the Izmir bay. In addition to the highway, a rail transportation system will also be included into the project.

The total length of the fixed link between Çiğli and İnciraltı regions of the Izmir bay is 6.8 km. The fixed link consists of a 4.2 km long bridge, 1.8 km long immersed tube tunnel and an artificial island in between. Immersed tube tunnel will be constructed under the current navigation channel of the Izmir bay. The shallow regions of the bay where the water depths are around 10 m will be crossed by the bay bridge. The location plan of the project is given in Fig. 1.

B. Kurtman (✉)
Bridge Department, Yüksel Project International, Ankara, Turkey
e-mail: bkurtman@yukselproje.com.tr

P. Gülkan et al. (eds.), *Developments in International Bridge Engineering*, Springer Tracts on Transportation and Traffic 17, https://doi.org/10.1007/978-3-030-59169-4_11

Fig. 1 Location plan

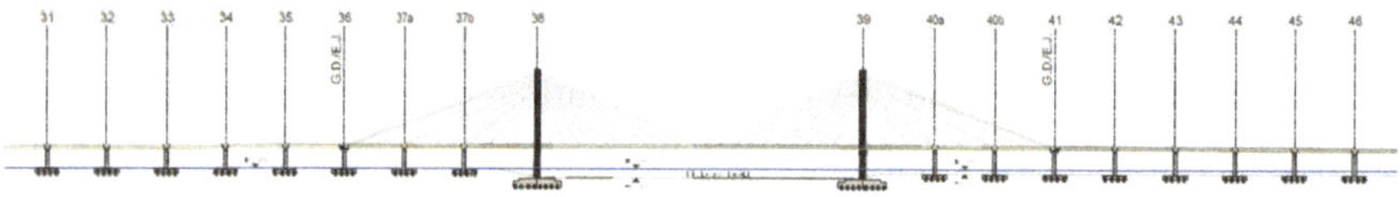

Fig. 2 Elevation view of the bridge (central part)

2 Description of the Bridge

Izmir Bay Crossing Bridge is a 4.18 km long bridge which consists of a 590 m long cable stayed central part and two approach bridges. North and south approach bridges have total lengths of 1735 m and 1850 m respectively. Post-tensioned concrete double cell box girder deck is used for the typical spans of 50 m. Cable stayed bridge has a 270 m concrete main span and 160 m side spans with intermediate piers. Deck is composed of two separate 21 m width platforms and is carrying three lanes of highway traffic and a light rail track on both platforms. Deck is connected to piers monolithically and at every 400 m there is an expansion joint. ϕ2000 mm diameter steel driven piles are used for the foundations. Each foundation is supported by 16 piles which have lengths between 70 to 86 m (Figs. 2, 3, 4 and 5).

Fig. 3 3D view of the bridge (central part)

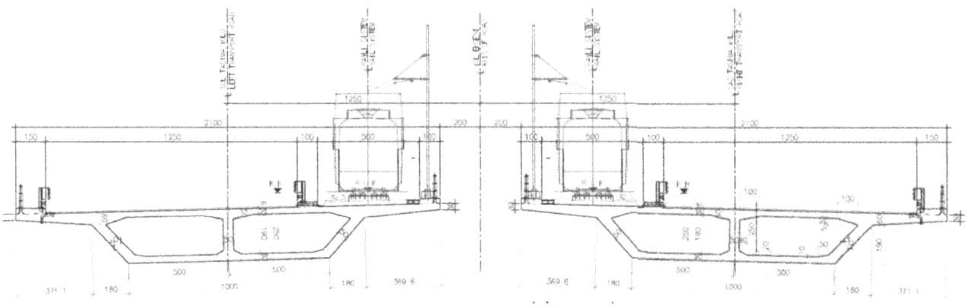

Fig. 4 Typical cross section of the deck

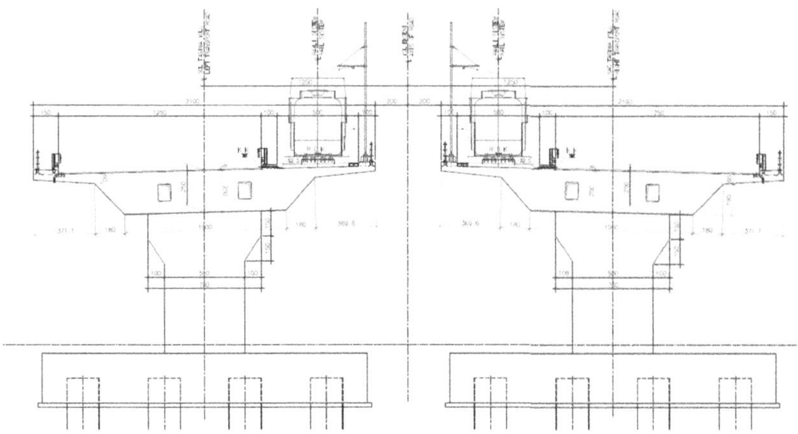

Fig. 5 Typical section of the piers

3 Design Criteria

3.1 Specifications

- KGM, Izmir Bay Crossing Special Technical Specification
- AASHTO LRFD Bridge Design Specifications, 2014
- AASHTO Guide Spec. for LRFD Seismic Bridge Design, 2011
- CALTRANS Seismic Design Criteria, 17th ed.

3.2 Materials

Concrete:

Post-tensioned deck and pylon : C50
Piers and foundation: C40

Post Tensioning Tendons and Stay Cables:
Grade 1860, Ø15.7 mm (Low Relaxation), 7 wire (pr EN 10138-3)

Reinforcement Bars: Pylon: S500, Other: S420.

Structural Steel: S355 (EN 10,025).

3.3 Live Loads

Standard H30-S24 highway truck loading is used as the design load. For rail loads three different train loads (tramway, Izban commuter and metro) are taken into account. Both highway and railway loads are applied to the deck simultaneously (Fig. 6).

3.4 Seismic Loads and Performance Criteria

Izmir bay is a highly seismic region with very poor soil conditions. Izmir and its surroundings have been the scene of intense earthquake activity since historical periods. A high seismicity is observed along the active faults in the region Alluvial deposits are variable thickness, about 100 m in the middle of the Bornova area and increase towards the bay. According to the results of the drillings the depth of the base rock in gulf area varies from 60 to 300 m. The marine alluviums observed in

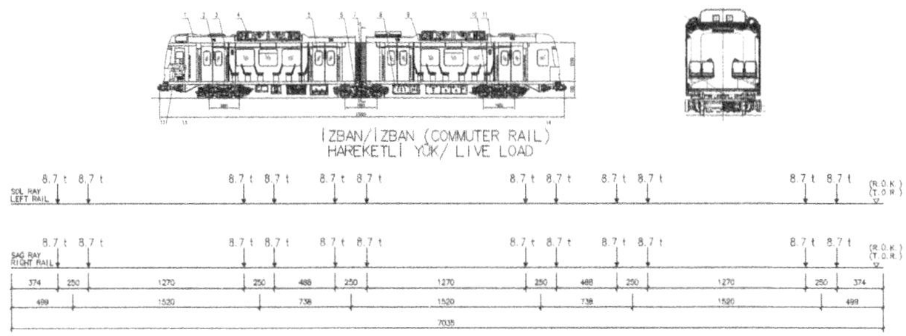

Fig. 6 Rail system loads

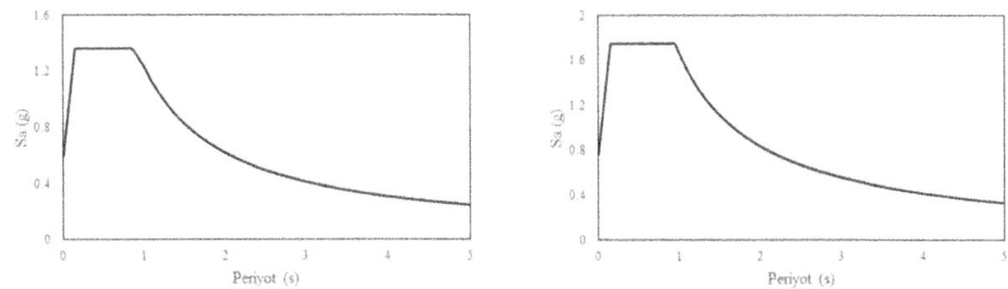

Fig. 7 Design spectrums (Tr = 1000 yrs and 2475 yrs)

Izmir Gulf are mainly composed of soft to very soft clay with included sand lenses. The design spectrums for NEHRP site class E are given below (Fig. 7).

The two-level design criteria approach is utilized to ensure that the Izmir Bay bridge which is constructed in a highly seismic area represents functional adequacy and economy while reducing life-threatening failure. The performance criteria is determined as follows:

For Design EQ (D1,Tr = 1000 yrs); Controlled Damage Performance Level.

This performance level corresponds to a state where non-extensive, repairable damage occurs in structures and/or in their elements under an earthquake. In this case, short-term (few weeks or months) interruptions in related operations may be expected.

For Design EQ (D2,Tr = 2475 yrs); Life Safety Performance Level.

This performance level corresponds to a state where extensive damage is expected in structures. But for this project; seismic joints used in the immersed tunnel and isolator bearings used in the bridge are not allowed to be damaged.

4 Analysis and Design

A 3D finite element model of the bridge was built and analyzed using Midas Civil program with post tensioning tendons in all construction stages accordingly. Static

Fig. 8 Structural models of approach bridge and cable stayed bridge

analysis, construction stage analysis with time dependent material properties, moving load analysis, unknown load factor calculations for cable stayed bridge and multi-mode spectrum analysis are conducted. Verifications for serviceability and ultimate limit states are performed in accordance with the rules set out in AASHTO LRFD Bridge Design Specifications (Fig. 8).

4.1 Determination of Final Geometry of the Cable Stayed Bridge

Cable stayed bridge geometry should be close to road geometry after the completion of the deck with superimposed loads. Also distribution of the bending moments due to permanent loads in the deck should be close to a beam elastically supported at the cable anchor points. This requires the determination and adjustment of the initial cable forces. The "unknown load factor" and "cable force tuning" functions in Midas Civil help to calculate initial cable tension forces while satisfying the specified constraint conditions of zero and the range of maximum and minimum values for displacements, reactions, member forces, etc (Fig. 9).

4.2 Live Load Analysis

Moving load analysis is conducted considering both highway and railway loads by defining the lanes, vehicles and loading scenarios. Worst positions of the vehicles for the maximum load effects are calculated by influence line method (Fig. 10).

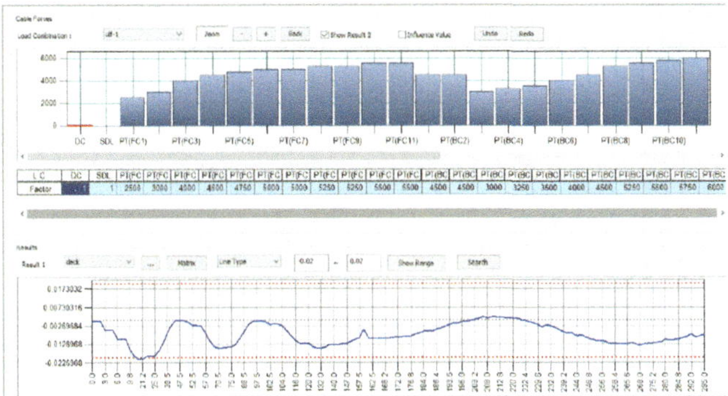

Fig. 9 Unknown load factor and cable tuning calculations

Fig. 10 Moving load analysis

4.3 *Seismic Design*

In order to take into account the soil-structure interaction in the analysis, the horizontal and rotational stiffness of the pile foundation are calculated by nonlinear pushover analysis. The soil is defined by nonlinear p-y curves and pushover curves are obtained by pushing the isolated pile foundation model for each degree of freedom. Then the capacity curves are used to determine the stiffness of the foundation and these values are defined in the global model. Also pile cap mass is lumped at the foundation level. As a result, the actual vibration frequencies are calculated. Since the first 20 m from the seabed is very weak clay with SPT values ranging from 0 to 5, the lateral resistance from soil is assumed to be zero (Fig. 11).

For approach bridge, the fundamental vibration periods for longitudinal and lateral directions are calculated around 2 s. Longitudinal displacements are calculated as 0.75 and 0.50 m at the deck and foundation level respectively. It is seen that implementation of mass and stiffness of the foundation makes significant contribution to the deformations of the bridge.

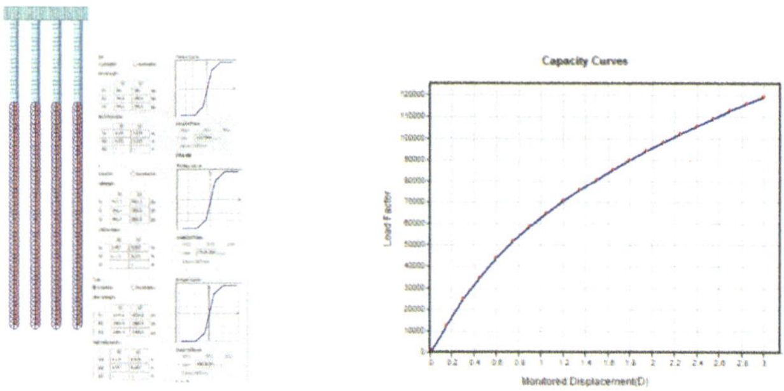

Fig. 11 Pile foundation modelling for pushover analysis

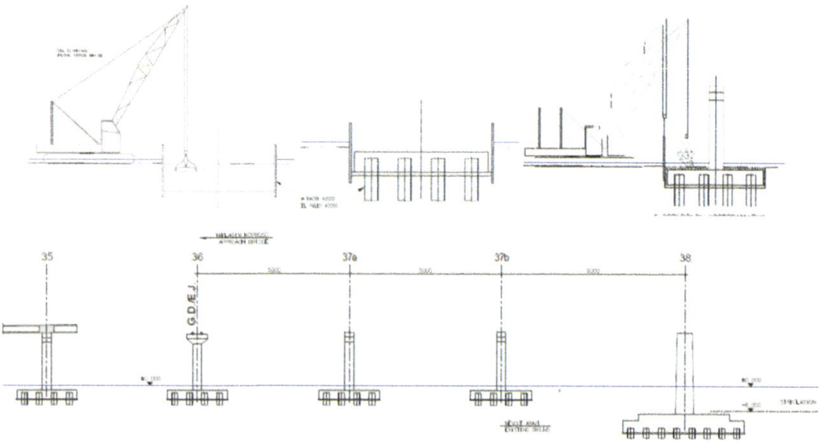

Fig. 12 Construction stages of substructure

5 Construction Method

During the final design stage, the construction method alternatives were studied by considering the site conditions, length of the bridge, time and cost. For the approach bridge which is 7170 m long with two separate platforms, movable scaffolding system is determined to be the most suitable construction method.

The construction of decks with several spans with movable scaffolding systems (MSS) may be a very efficient and competitive constructive method. This solution is generally used for the 40–60 m span range. Moreover, the use of MSS may represent very significant cost reductions if the access to the front line of the site is difficult – for example, high piers or water access, because this may imply for significant cost of elevation equipments. It is possible to complete a span around 5 days in this method. As a result, it is considered as the most advantageous method for the Izmir bay bridge which has 144 spans in total (except for the cable stayed part) (Fig. 12).

The construction method for the main span of the cable stayed part is selected as the cantilever method which is the most common method for cable supported decks. The side spans of the cable stayed bridge will be completed by MSS method. After the completion of the side spans and pylon elevation, main span segments will be cast in-situ by form traveller (Fig. 13).

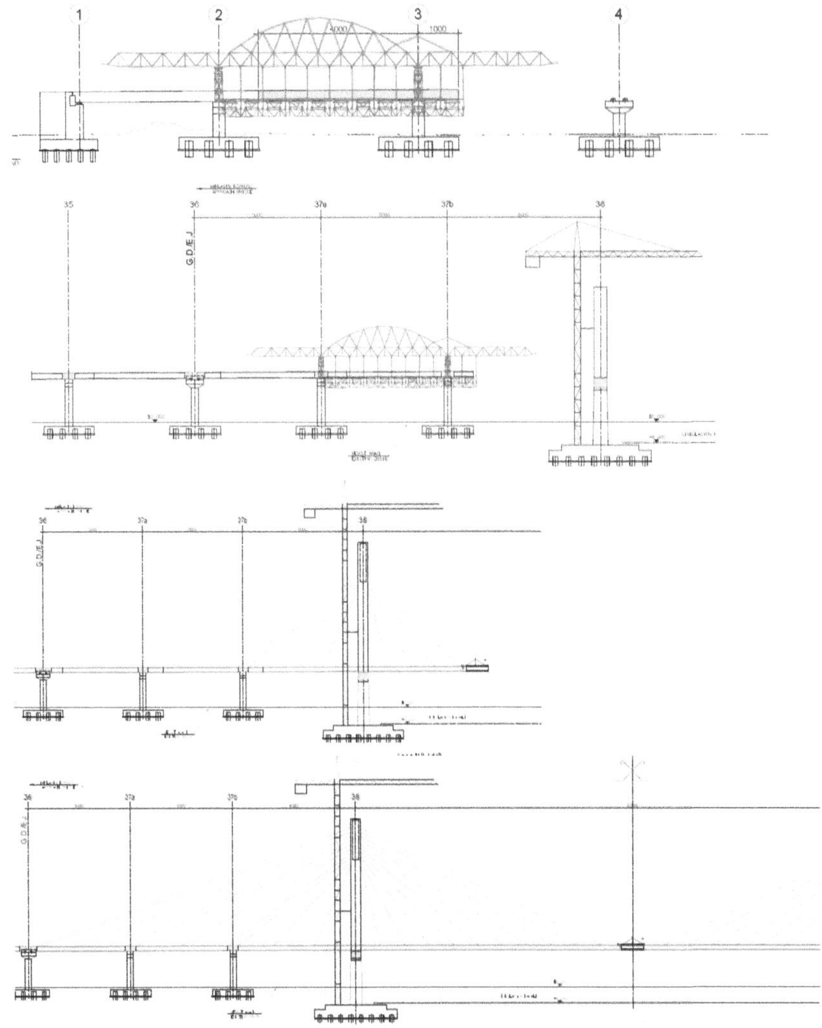

Fig. 13 Construction stages of deck

6 Conclusion

Designing a long water crossing bridge in very poor soil conditions and high seismic region is a challenging task. Standard span lengths and modern construction technologies make this bridge feasible. In poor soil conditions, it is necessary to include the foundation stiffness and mass in the analysis. This can be done by modelling all the substructure elements in the global model or by defining the lumped stiffness and mass properties in the global model. The stiffness properties are calculated from a separate model, which a pushover analysis is performed with nonlinear soil springs. As a result, realistic vibration frequencies and deformations are calculated.

Acknowledgements This document was prepared for the Istanbul Bridge Conference 2018.

Bridge Construction

Al Bustan South Bridge Design and Build Project, Doha, Qatar

Yousef Al Emadi and Ali Kara

Abstract This paper summarizes recent developments in the design and built Al Bustan Bridge and Highway Project as part of the Qatar Expressway Programme and the lessons learned from it.

The Al Bustan South Project comprises design and construction of an expressway to the north of Doha with grade separated junctions, four-lanes dual deck to form a 1,946 m long bridge structure, including 863 on/off ramp bridges. The project includes drainage system, extra high-voltage cables, street lighting, substations, irrigation transmission pipelines, water pipelines, foul sewer, landscape, intelligent transportation, art-scaping design and construction.

The Qatar Expressway Programme is a very large road infrastructure projects that connects Doha with other cities through a set of advanced highways, roads, and flyovers. The scope is expected to deliver about 800 km of safe and efficient roads. The programme also includes the construction and development of an integrated infrastructure network equipped with state-of-the-art, reliable underground utilities, including storm water networks, electrical services, and intelligent transport systems. Al Bustan Corridor, Orbital Highway, North Road, Al Rayyan Road and Lusail Expressway part of this programme.

Design and building of highways and bridges in urban areas with very tight schedules are always big challenges to contractors as well as to the employers. The contractor must manage the busy existing traffic to provide work zones to the bridge construction works. The existing utilities and their relocation have been major issues in the structural design to accommodate the foundations. Consideration of utility relocations and traffic diversion management may change the entire concept of bridge's design and construction. To improve life around the city and achieve aesthetic harmony in these structures, the employer had to make important provisions.

Y. Al Emadi · A. Kara (✉)
Public Works Authority, Doha, State of Qatar
e-mail: akara@ashghal.gov.qa

P. Gülkan et al. (eds.), *Developments in International Bridge Engineering*, Springer Tracts on Transportation and Traffic 17, https://doi.org/10.1007/978-3-030-59169-4_12

1 Introduction

Public Works Authority, Qatar Expressway Programme is the World's largest road and infrastructure Programme brings together around 60 road projects, 800 km of safe efficient roads, and 240 major interchanges.

Al Bustan Corridor is very important alternative to existing expressway named 22nd February corridor in national expressway which crosses free flow traffic length 14 km from North of Central Doha to South consists of 4 Contracts. Moreover, Al Bustan South project is one of the contracts which was previously only a built scheme with a (−1) level underpass structure on mainline section. During the tender stage of the built only scheme, Public Works Authority issued an addendum to the Tenderers to propose an alternative design-built for mainline alignment in order to reduce the construction cost and to have sustainable structure design. The project was then awarded on this basis as design-built with the detailed design to be resubmitted on the basis of the (+1) bridge structure. Project location is shown in Figs. 1 and 2.

2 Overview of Scope

Al Bustan project upgraded to a dual carriageway with minimum of 4 lanes in each direction from South of Rayyan Road Junction R6 to South of Al Waab Street (Fig. 3). Within the project limit the mainline Al Bustan Street at Junction B1, where Al Waab Street intersects with Al Bustan Street: a three level interchange shall be constructed with a signalized at-grade junction. At Junction B2 where Al Khufoos

Fig. 1 State of Qata

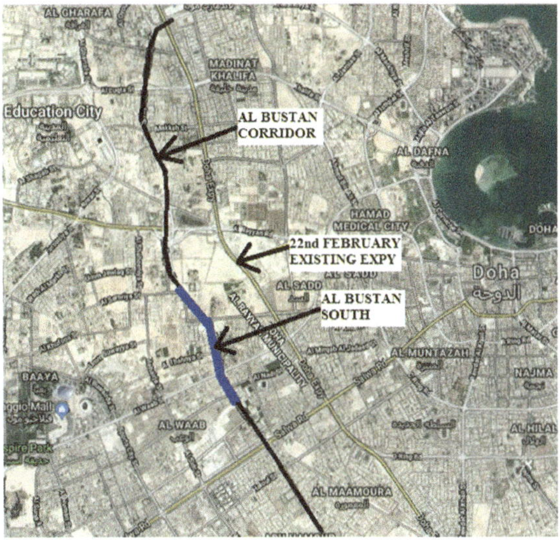

Fig. 2 Al Bustan Corridor and Al Bustan South

Fig. 3 Al Bustan South bridge and junctions B1, B2, R6

Street (Al Rasheeda) intersects with Al Bustan Street: a two level interchange shall be constructed with a signalized at grade.

Furthermore to maintain East–West local accessibility, a mid block crossing underpass shall be constructed between junction R6 and Junction B2 (Fig. 3 and 4). The total length of the improvement scheme is approximately 5 km with associated service roads, minor junctions and structures. Project Estimated cost 1.58 bn

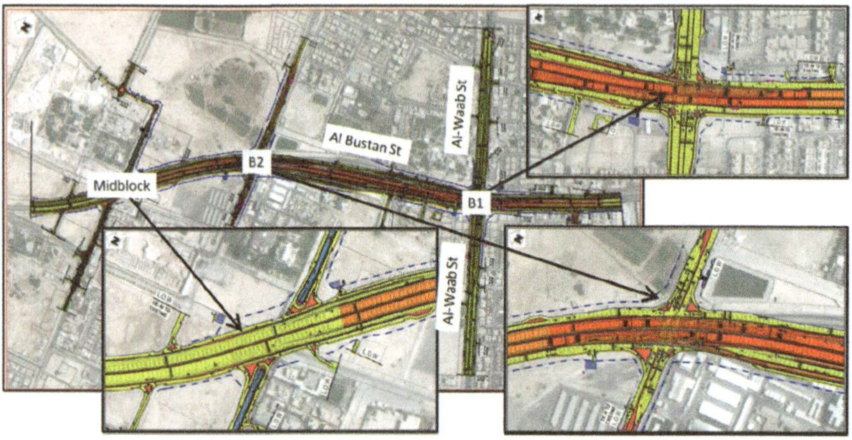

Fig. 4 Al Bustan South interchange locations

Qatar Riyal ($434 Million) and Construction duration from October 2017 to January 2020.

3 Design Considerations

3.1 Highway Details

Al Bustan Street is classified as an Urban Expressway. It is out of the Classified High Load Route of Qatar. Permitted Traffic Speed as following;

Location/Link	Design speed (Kph)	Posted speed (Kph)
Al Bustan street (main line)	100	80
Ramps	70	60
Al Waab street	100	80
Al Rasheeda street	70	60

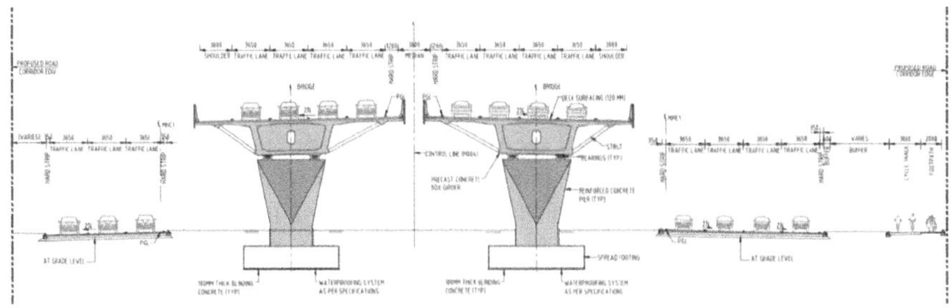

Fig. 5 Al Bustan South corridor highway cross section

Table 1 Cross section configuration

Location/Link	Proposed cross section
At grade road, north bound	Three 3.65 m through lanes
North bound bridge mainline carriageway	3.0 m outer shoulder, four 3.65 m traffic lanes, 1.2 m inner shoulder (hard strip)
Centre median	3.0 m, separated by F shape concrete barriers
South bound bridge mainline carriageway	3.0 m outer shoulder, four 3.65 m traffic lanes, 1.2 m inner shoulder
At grade road, South bound	Three 3.65 m through lanes

3.2 *Highway Cross Section*

The highway cross section on the Al Bustan street bridges comprises four lanes of traffic in each direction. A typical highway cross section is shown in Fig. 5 and dimensions given in Table 1.

3.3 *Description of Bridge Structures*

The bridges are twin parallel structures, each catering for one carriageway of Al Bustan Street. These bridges cross Al Waab Street (Junction B1) and Al Rasheeda Street (Junction B2) intersections shown in Fig. 4. The overall length of each bridge is 1897.2 m between abutments with 34 no. of spans comprising typical spans of 57 m and maximum span of 72 m at intersections.

Ramp connections are proposed at four locations; Ramp bridges are three to five span viaducts separated by expansion joints. Total length of four ramps is 863 m with typical spans of 57 m.

3.4 Structure Type

The superstructures for Main Bridges BR-1A and BR-1B are parallel structures, each of them are single cell precast segmental box girder of depth 3.20 m for typical spans of 57 m, construction by Balanced Cantilever Construction Method (BCM) and for the varying width deck consists of two cell post-tensioned box girder, with cast-in-situ construction. At the junction location depth of box girder varies from 3.2 m to 4.0 m for 72 m spans. The superstructure width of typical cross section is 19.8 m, which is varying from 19.8 m to 29.6 m at the four ramp connections. The box girder cantilevers are supported by reinforced concrete struts at 1.5 m spacing along the bridge.

The main bridges comprise segments/modules separated at expansion joint to accommodate longitudinal movement of deck due to temperature variation. Each module consists of multiple spans continuous over pier and resting on mechanical bearings at pier and abutment.

The bridge deck section for ramps is single cell post-tensioned box girder of 3.2 m depth and an overall width of 8.8 m, construction by precast Balanced Cantilever Construction Method (BCM).

The abutments for both main and ramp bridges are wall-type abutments retaining the backfill behind the wall. The Piers for both main and ramp bridges are single tapered reinforced concrete piers with mechanical bearings.

All structures of main bridges and ramps are designed in accordance with the 'Design Criteria for Highway Structures Interim Advice Note 009 (IAN009), Revision A2' and the requirements for bearing and expansion joints IAN 006 and IAN 031 respectively produced by Public Works Authority. As per the criteria, design carried out using relevant British Standards, Design Manual for Roads and Bridges and UK's Highway Agency Technical Memoranda. Seismic loading designed in accordance with AASHTO LRFD 2014 with a peak ground acceleration of 0.15 g (Fig. 6).

3.5 Foundation Type

Foundation of piers and abutments comprised of isolated spread footings and pile groups mostly on rock strata.

3.6 Articulation Arrangements

For main bridges, the deck is formed by one single box girder for typical spans with constant 19.8 m wide deck and by two cell box girder at junction locations including at ramp connections where the roadway width varies from 19.8 to 29.6 m. The box girders will be supported on two mechanical bearings at each column location and

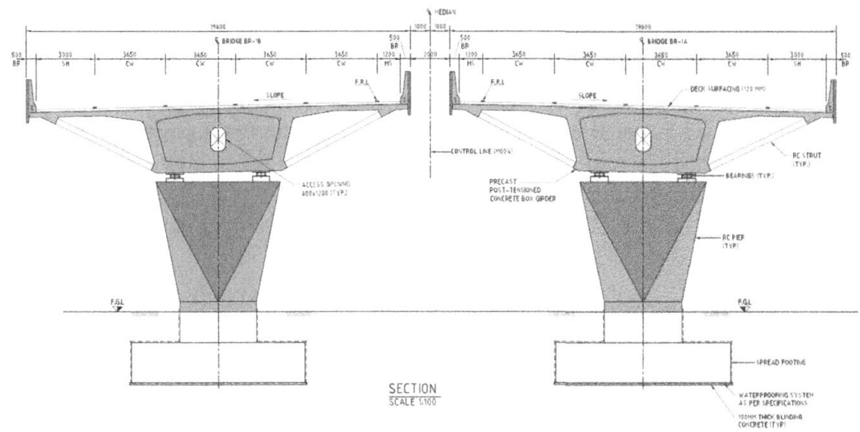

Fig. 6 Bridge cross section

at abutments for typical spans and three mechanical bearings where the deck width varies. For main bridges, in addition to the expansion joints at end abutments there will be 6 intermediate expansion joints to form the main bridge with 4 to 6 continuous spans for full length of the bridge as 7 units.

The basic articulation principles are for each of the 7 segments of the main bridge, the deck will be both longitudinally and transversely fixed at one internal column (central or near central) by monolithic connection.

All other piers including portals, expansion joint piers and abutments, have bearing articulation with two mechanical bearings with one longitudinally guided bearing to resist transverse forces and one free bearing to allow for both transverse and longitudinal movements.

Typical bearings articulation plan is shown below;

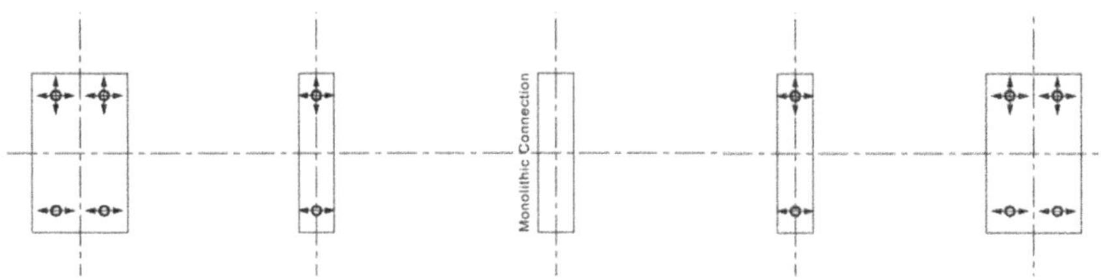

Typical bearing articulation - plan

Ramp bridges follow a similar approach and supported on two mechanical bearings at each column location and at abutments. The longitudinal fixation of the deck located at one internal pier (central or near central) of each ramp bridge. The longitudinal fixation made through a monolithic connection. The fixed pier designed for all longitudinal traffic loads and horizontal seismic loads. One multi-directional bearing and one guided bearing—restraining the deck transversally—located at all other columns and abutments.

3.7 Road Restraint Type

Parapets on the bridge deck are F shape reinforced concrete precast parapet with H4a containment level as per BS EN 1317–2:2010 to meet the Expressway Programme criteria for bridges. The minimum height of parapet with H4a containment as required by the expressway programme criteria is 1500 mm.

3.8 Design and Service Life

The overall service design life of the bridge structure are 120 years and shall represent the assumed period for which a structure or part of it is to be used for its intended purpose with anticipated maintenance but without major repair being necessary.

Design lives are design service life (years) for mechanically Stabilized Earth Walls 120 years,

Non-replaceable expansion joint components 50 years, bearing devices 50 years, Asphalt wearing course 20 years.

3.9 Materials

3.9.1 Structural Concrete Classes

The classes of concrete are based on 28-day minimum compressive cube strength, which is estimated to be 1.25 times the cylinder strength. The classes of concrete is denoted as cube strength/maximum aggregate size (MPa/mm) and specified for use as follow:

Pre-stressed precast concrete girders	C60/20 (48 MPa cylinder)
Cast-in-place post-tensioned and reinforced concrete deck	C50/20 (40 MPa cylinder)
Reinforced concrete in tunnels, underpasses & retaining wall	C40/20 (32 MPa cylinder)
Approach slab, abutments, foundations, piles and pile caps C40/20	(32 MPa cylinder)
Piers C50/20	(40 MPa cylinder)
Traffic barrier, precast MSE wall panels	C50/20 (40 MPa cylinder)
Blinding and unreinforced concrete	C30/20 (24 MPa cylinder)

3.9.2 Reinforcement

All reinforcement are deformed round steel bars Grade B500B or B500C conforming to BS EN 10,080 and BS4449: 2005 and cut and bent in accordance with BS 8666:

2005. The characteristic strength of reinforcement is 500 N/mm^2 with a minimum modulus of elasticity of 190 kN/mm^2. Due to the significant inertia forces, a ductile design for built in piers has been adopted with a maximum value of 2 for the response modification factor. AASHTO detailing rules followed including the provision of closed stirrups and splicing of vertical rebars outside plastic hinge zone.

3.9.3 Prestressing Strands

Prestressing strands are uncoated, seven wire, super low relaxation strands conforming to Part 3 of BS5896.

The following tendon properties used in design:

- Seven wire (super) low relaxation strands, Strand relaxation of 2.5%, Nominal diameter of 15.70 mm, Steel cross sectional area of 150 mm^2, Tensile strength of 1860 N/mm^2, Guaranteed ultimate tensile strength of 279 kN per strand, Modulus of elasticity of 195 kN/mm^2, the following parameters used in design:
- Friction wobble coefficient, K, of 0.0017/m, Friction curvature coefficient, μ, of 0.14/radian., Anchor set of 6 mm., Relative humidity of 70%.

The following limiting values used for the design of post-tensioned tendons:

- Maximum jacking force equal to 80% of guaranteed ultimate tensile strength, Maximum force immediately after anchor set at anchor equal to 70% of guaranteed ultimate tensile strength.

Structures with external and unbonded prestressing comply with the requirements of BD58 and the guidance in BA58. Structures with external and unbonded prestressing designed with provision for the replacement and re-stressing of tendons. The structure capable of accommodating a range of prestressing force resulting from future tendon replacement, and galleries are incorporated at abutments and intermediate piers expansion joints with sufficient room to manoeuvre prestressing jacks. Concrete boxes housing external tendons have adequate working space and means of access to facilitate replacement and restressing of tendons. The proposed stressing arrangement for external tendons has been done considering the above provision and all live end stressing has been proposed from inside the box superstructure. Anchorage end at expansion joints will act as dead end anchorage; sufficient space has been provided at abutment and intermediate pier expansion joint locations for installation, inspection and maintenance of these dead end anchorages.

Structures with external and unbonded prestressing designed to permit future dead load or deflection adjustment as outlined in Clause 5.14.2.3.8.c of AASHTO LRFD 2014. This consist of provision for the future addition of external post tensioning located inside the box structure and symmetrical about the structure center line. Anchorages and deviators provided in addition to all necessary openings through concrete elements. These have the capacity to allow an additional prestressing force to be introduced in the future of not less than 10% of the positive moment and negative moment prestressing force provided at the time of construction.

Ducts for unbonded tendons within the concrete section are made from High Density Polyethylene (HDPE) with a minimum wall thickness of 2 mm. Ducts waterproofed with drainage holes at low points along their profile to release any water that may be trapped and to facilitate inspection by means of an endoscope.

Unbonded tendons within ducts are corrosion protected with galvanising, painted in a light colour to aid inspection which allows for tendon replacement and restressing.

3.9.4 Prestressing Ducts

Prestressing ducts consist of an electrically non-conductive and corrosion resistant durable material such as high density polyethylene or polypropylene forming a double corrosion protection system in combination with the grout. Ducts supported at a maximum spacing of 1000 mm.

3.9.5 Grout for Post-tensioning Tendons

Grout for post-tensioning tendons conform to the following requirements:

- The compressive strength of 100 mm cubes are made of the grout shall exceed 27 N/mm^2 at 7 days and at least 62 N/mm^2 after 28 days. Cubes made, cured and tested in accordance with BS 1881.
- Grout shall consist only of Portland cement complying with BS12 Class 42.5 N, admixtures complying with BS 5075:

  Parts 1 and 3 and water complying with BS 3148.
- Chloride ion content of the grout shall not exceed 0.1% chlorides by mass of the cement.

3.9.6 Bearing Devices

Bearing design, manufacture and installation comply with the requirements of BS5400: Part 9: 1983 and BD 20/92. All bearings are replaceable, and positively located to prevent them from shifting during bridge movement. Adaptor plates are provided at the top and the bottom of the bearing device to facilitate simple removal. Bearing types which are acceptable for installation at bridge supports are proprietary mechanical bearings such as mechanical bearings. Bearings protected by removable aluminum dust cover plates, and located in the manner that exposure to standing water is prevented.

3.9.7 Expansion Joints

Expansion joints are in accordance of the requirements of EOTA for modular expansion joints and BD 33/94. Modular type expansion joints are adopted to accommodate higher movement range for all bridges. Provisions for adequate space between adjacent deck faces have been made to unable underside inspection of joints.

4 Geotechnical Summary

The ground materials and material properties encountered during the Al Bustan Street South site investigations are typical of the Doha geology and no onerous or high risk geotechnical materials have been encountered. In addition, detailed design Geophysical investigations indicate that there is a low risk of subsurface cavities and voids at the Al Bustan Street South site. A generalized ground profile encountered along the project site is included in below.

Generalized ground model from Al Bustan South stratigraphic unit	CODE	Average top elevation (M QNHD)	Average bottom elevation (M QNHD)	Average thickness (M)
Made ground	**FILL**	+10.20	+9.55	0.65
Residual soil	**RES**	+9.60	+8.90	0.70
Simsima limestone C	**SLC**	+8.90	+1.90	7.0
Simsima limestone B	**SLB**	+3.30	−2.60	5.9
Simsima limestone A	**SLA**	+2.20	−4.10	6.3
Midra shale	**MSH**	−4.10	−7.30	3.2
Rus formation	**RUS**	−7.40	−16.5	>9.1

QHND: Qatar National Height Data.

5 Methods of Analysis Proposed for Superstructure, Substructure and Foundations

Global Analysis.

The bridges analyzed using independent models for superstructure and substructure:

5.1 Superstructure

Pre-cast segmental deck modeled using commercial software MIDAS Civil, a 3-dimensional finite element (FE) software that models accurately also effects of prestressing, creep, shrinkage and temperature. Two different models used for the analysis of the deck:

1. Longitudinal model: The model reflect the articulation of the bridge and will take into consideration the erection methodology and the precast segmental construction history. The deck units modelled using beam elements. Long term effects of the concrete are taken into account considering the estimated casting and erection date of each individual segment. Longitudinal models for the ramps and longitudinal models for the main bridges built to calculate the longitudinal stresses. The columns modeled using linear elements restrained at foundation level.
2. Transverse model. The model a 3D FE model using shell elements and limited to one span from column to column including end diaphragms.

A 3D FE shell model built for single box bridges. The effects of the transverse model combined to the effects of the longitudinal models for compatible load combinations.

5.2 Substructure

A different calculation model from the one used for the Superstructure, considering the effects of the seismic load defined in following paragraphs in this paper,

Vehicular live load surcharge considered in the design in accordance with Clause 5.8.2 of BD 37/01 as HA Loading: 10kN/m^2, HB 30 units: 12kN/m^2, HB 45 units: 20kN/m^2.

Live loads considered in accordance with Sect. 6.3 of IAN009 and BD 37/01, BD 24/92.

Wind load determined in accordance with the requirements of IAN009,

Site Hourly Mean Wind Speed, Vs: 26 m/s,

Maximum Wind Gust Speed, Vd: 45 m/s for bridge without live load,

Vd: 35 m/s for bridge with live load,

Nominal Transverse Wind Load, Pt: $0.613Vd^2\ A_1\ C_D$ (N), where A_1 = the solid area (m^2), C_D = drag coefficient for bridges in the gulf area with C_D max ≤ 2.

The wind speeds are appropriate to a height above ground level of 10 m in open level country with an annual probability of being exceeded of 0.02 (50-yr return period).

Dead and superimposed dead loads:

Reinforced or Prestressed Concrete: 25.0 kN/m^3.

Unreinforced (Mass) Concrete: 24.0 kN/m^3,

Structural Steel: 78.5 kN/m^3,
Wearing Surface: 23.0 kN/m^3,
Initial Wearing Surface (120 mm minimum) 2.76 kN/m^2 curb to curb,
Future Additional Wearing Surface (50 mm thick): 1.15 kN/m^2 curb to curb,
Soil Backfill: 20 kN/m^3.

Utilities: 2 kN/m (No major utilities are proposed on the structure, however a provisional load of 2 kN/m is considered to cover lighting arrangements),

Bridge Barrier (1.5 m high F shape): 17.0, 0.5 kN/m allowance has been made for Aluminium cladding.

Global analysis of the structure performed using MIDAS Civil finite element analysis software. The superstructure idealized as spine beam and the piers/abutments connected to it by means of joints.

The Pier columns modeled as frame elements and connected to the superstructure with the joint conditions that reflect the articulation of the bridge bearings. Columns modeled up to the top of the foundation, connected to raft foundation as plate element with springs considering soil subgrade modulus.

For pile foundations, the ground condition/soil parameters introduced into the software by means of horizontal Winkler spring modulus as stated in the geotechnical interpretative report.

5.3 *Response Spectrum Analysis*

For extreme event, the seismic forces calculated by performing Response Spectrum Analysis (RSA) in accordance with AASHTO LRFD 2014, cl. 4.7.4.3.

The design based on Seismic Zone 1 and site class B, which confirmed in the geotechnical report. The substructure designed to resist the earthquake forces calculated by rigorous analysis (response spectrum analysis) as considering the parameters: PGA = 0.15, As = 0.15, SDS = 0.375, SD1 = 0.15.

For Piers, foundations, and connections, design forces obtained from the analysis (RSA) modified using response modification factor R corresponding to bridge operational category "Essential". For single piers (wall type), response modification factor of 2 in weak direction (longitudinal) and 1.5 in strong direction (transverse) is adopted. For portal piers, response modification factor of 2 is adopted in weak direction (i.e. transverse seismic), whereas for strong direction response modification factor is not applicable as the bearings are free in this direction (longitudinal).

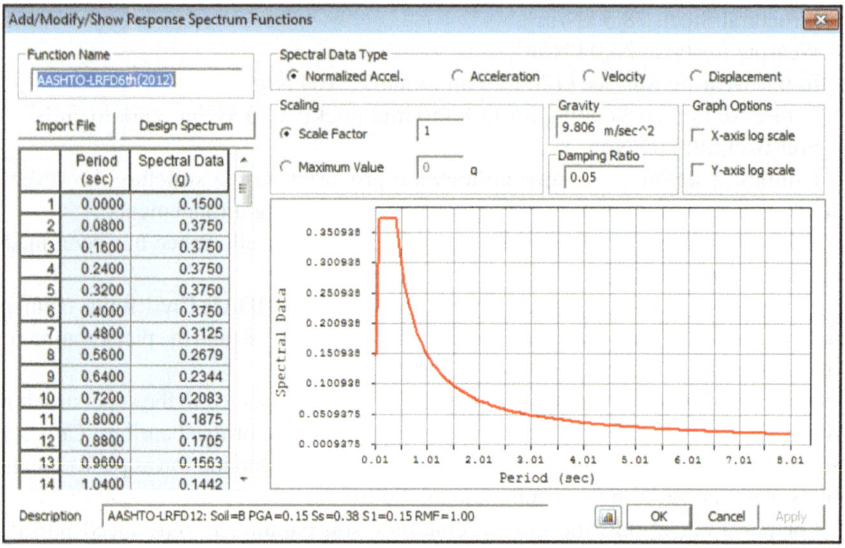

- 100% of the absolute value of the force effects in one of the perpendicular directions combined with 30% of the absolute value of the force effects in the second perpendicular direction.
- 100% of the absolute value of the force effects in the second perpendicular direction combined with 30% of the absolute value of the force effects in the first perpendicular direction.

The elastic seismic forces thus obtained from analysis (RSA) in the two perpendicular directions combined to form two load cases as follows:

Design forces for each structural component obtained by modifying the elastic forces calculated from the above two load cases using the appropriate response modification factors R as given above. Reinforcement detailing of substructure as per the seismic design provisions given in clause 5.10.11 of AASHTO LRFD 2014.

Stability of foundations under seismic loading checked in accordance with AASHTO LRFD 2014.

6 Conclusions

Al Bustan South Project procurement stage opened to the Tenderers as to be only construction project with complete and approved design as (−1) level underpasses. However, during the Tender stage the Public Works Authority requested an alternative cost reduced design proposal from the Tenderers providing the same mobility, accessebility and functional design as to be Design-Built contract. A Contractor proposed an alternative design-built as to be (+1) bridge instead of (−1) underpass design and the Procurement stage was ended with succes to award the project as

Design-Built with significant cost saving to the Public Works Authority. The above information is given to demostrate the bridge design criteria along with methods of analysis and appropriate design approach for sustainable bridge construction. Considering state of art expansion joints and bearings standing extreme weather conditions, additional post tensioing ducts for future use, special care for inspection to expansion joints, bearings and post tensioning, highest containment level F shape concrete parapet, designing the deck as single cell precast segmental box girder for typical spans construction by Balanced Cantilever Construction Method and for the varying width deck consists of two cell post-tensioned box girder, with cast-in-situ construction successfully achieved.

Acknowledgements The Authors wish to express their gratitude to Dr. Eng. Saad bin Ahmad Al Muhannadi for his invaluable support on the preparation of this article.

Design Challenges of a River Crossing: Dim Çayı Extradosed Bridge

Cemal Noyan Özel, Kamil Ergüner, Abdullah Rahman, Sema Melek Kasapgil, Hatice Karayiğit, Özgür Özkul, and Kutay Kutsal

Abstract Extradosed bridges are efficient and aesthetic alternatives for relatively long urban crossings with shallow decks. The first extradosed bridge in Turkey was constructed in Antalya Çallı Crossing, with an 80 m main span and 2.5 m deep post-tensioned pi-section deck. The second extradosed bridge is currently under construction over the Dim River in Alanya, as part of the new Alanya-Gazipaşa Highway. The Dim Çayı Extradosed Bridge has seven spans and consists of two parallel decks: 400 and 390 m long, with an S-shaped curve in plan. 12 m high pylons, housing two sets of six continuous cables per pylon, using the Freyssinet Saddle and Cohestrand technology, support the 100 m main span. The variable height post-tensioned pi-section deck is cast-in-place over scaffolding. Couplers and external blisters provide the continuity of the tendons. Construction stage planning of the bridge required the casting of the first half of the viaduct, and resting the 50 m half mid-span on temporary steel piers. River bed is temporarily diverted to base the temporary steel piers and its foundation. The construction of the concrete pylons and saddles are followed by stressing of the extradosed cables and removal of temporary piers. Length of the bridge, curvature in plan, variable deck geometry, number of cables and the location of riverbed, created challenges on the construction stage analysis. The deflection of the deck is monitored to check it stays within the tolerances of the theoretical results. For the seismic design, a return period of 1000 years was taken. Seismic acceleration and short piers required the use of a combination of Lead Rubber Bearings and Pot Bearings to reduce the seismic effects and lead to an economic and safe substructure solution. At time of preparing this paper, the bridge was under construction and was to allow traffic by the end of 2018.

C. N. Özel (✉) · H. Karayiğit · Ö. Özkul · K. Kutsal
Freysaş-Freyssinet Yapı Sistemleri A.Ş., Istanbul, Turkey
e-mail: cemalnoyan.ozel@freysas.com.tr

K. Ergüner · A. Rahman · S. M. Kasapgil
Structural Engineering Department, Mega Mühendislik Müşavirlik A.Ş., Çankaya/Ankara, Turkey

P. Gülkan et al. (eds.), *Developments in International Bridge Engineering*, Springer Tracts on Transportation and Traffic 17, https://doi.org/10.1007/978-3-030-59169-4_13

1 Introduction

The importance of transportation increased due to the rapid urbanization in last decades. With the increase of transportation projects, various types of bridge construction methods have been developed considering the probable constraints in construction sites. The advances in technology brings evolution to new bridge types. Extradosed bridge is a relatively new bridge construction method.

Extradosed bridge can be defined as a combination of prestressed girder bridge and cable-stayed bridge. It is evolved from cantilever bridge concept with external tendons. Extradosed term comes from a French word "extradoses" which means upper or outer curve of an arch [1] (Fig. 1).

Extradosed bridge is a hybrid form of a stay cable and a girder bridge. Although the outer view looks like a stay cable bridge with a shorter pylon, static behavior of deck is more likely a post-tensioned deck with external tendons. Due to low inclination angle of the stay cables, they provide more axial force like post-tensioning instead of uplift force. In the literature, there is no certain border between stay cable and extradosed bridge but generally, pylon height of extradosed bridges is less than one eighth of main span, corresponding to a cable inclination about 17° [1]. Cable force that is on one side of the pylon needs to be transferred to the other side of the pylon. In ordinary stay cable bridges, generally cables are anchored both at the deck and at the pylon, with anchorages due to larger inclination angles. In this case, load can be transferred with overlapping of cables, continuity tendons or steel parts inside the pylons. Smaller cable angles of extradosed bridges can provide a simpler solution: saddles. Saddle allows the cables to pass through the pylons without a discontinuity. Saddle can be designed to transfer the differential force with friction or bonding. By using saddles, pylon dimensions can be reduced resulting to pylons that are more compact. Another main difference of extradosed bridges compared to stay cables is the deck stiffness. Extradosed bridges have an option to create stiffer deck (concrete deck) in order to resist live loads. Increase on deck stiffness compared to cable stayed bridges provides less live load stress range on cables. Since live load stress range in cable is smaller, well-known codes around the world allow 60% of guaranteed ultimate tensile strength (GUTS) of the extradosed cables, while stay cable bridge limit is 45% of GUTS under the service loads [3]. This leads to a more compact and efficient cable design for extradosed bridges.

Fig. 1 Kiso Gawa Bridge, Japan [2]

Extradosed bridges provide efficient solutions for the span lengths ranging from 80 to 250 m as an alternative to girder bridges. In Turkey, the very first extradosed bridge "Çallı Bridge" was built in Antalya city center in 2016. It is part of a three-level interchange project. In order to provide sufficient clearance under the deck with an iconic appearance, Çallı Bridge is built using the extradosed concept. The construction of the Çallı Bridge is completed in six months. Success in this project opened the door to a second one, which is a river crossing: the Dim Çayı Bridge.

2 Dim Çayı Extradosed Bridge

2.1 General Overview

Dim Çayı Bridge is located close to Alanya city center, in south Turkey as part of new Alanya-Gazipaşa ring road project. The bridge project consists of two parallel 13 m wide decks over Dim River. The road has three different alignment in plan as it is formed considering the river passage and the tunnels located behind the abutments. Bridge starts with an upward curve, continues with a straight part and ends with a downward curve in plan (Fig. 2). The clear spacing between decks is 16 m in plan. The left deck (north) is 400 m long, while the right one (south) is 390 m. In order to locate bridge piers out of the river bed, 100 m minimum main span length is needed. Extradosed bridge option was chosen from three different alternatives evaluating the riding comfort, construction economy, horizontal clearance and asthetical view. Project owner is KGM (General Directorate of Highways), and the designer is Mega Mühendislik Müşavirlik A.Ş. Main contractor is HGG İnşaat.

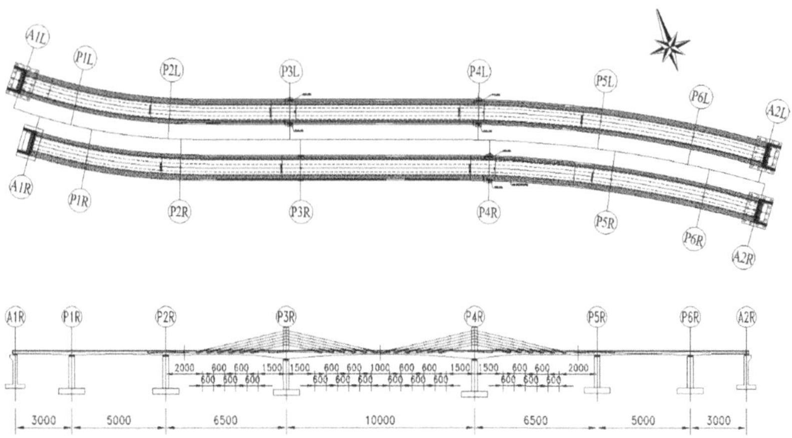

Fig. 2 Plan and elevation views of the bridge

Deck is designed as post-tensioned concrete with a pi-shape (double-tee). Deck height is variable, ranging from 4.4 m height (max) at pylon to 2 m height for typical mid-spans. Deck width is 13 m in both decks. At support and stay cable anchorage locations more rigid sections, namely "diaphragms", are created to resist unbalanced loads and torsional effects. In addition, these diaphragms help to distribute the load uniformly to the whole section. The bridge consist of seven spans, ranging from 30 m at side spans and 100 m at main span (Fig. 2). In longitudinal direction, in addition to the extradosed cables, 5 × 31c15 internal post-tensioning tendons are used in each rib, resulting to 10 × 31c15 tendons in each deck. Typical deck section at mid span is presented in Fig. 3.

12 m tall rectangular pylons at P3 and P4 axis on each deck are designed to deviate the cables. The pylon height corresponds to one eighth of the main span, which is typical for extradosed bridges. On each pylon, six layers of 31c15 stay cables are used (Fig. 4). The harped cable configuration consists of parallel cables anchored over the full height of the pylons, which provide similar cable angles and ease of construction.

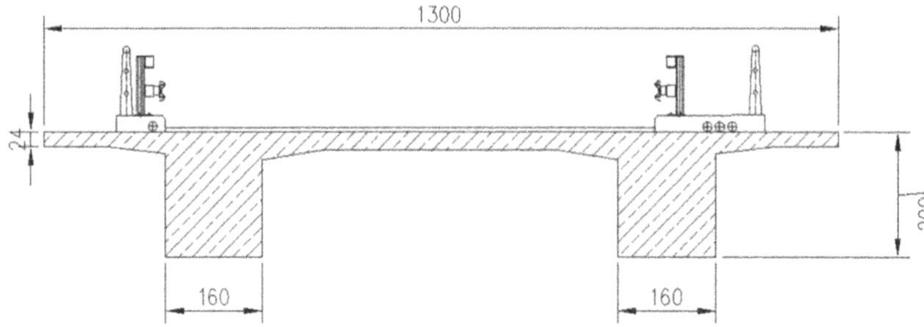

Fig. 3 Typical mid-span cross-section

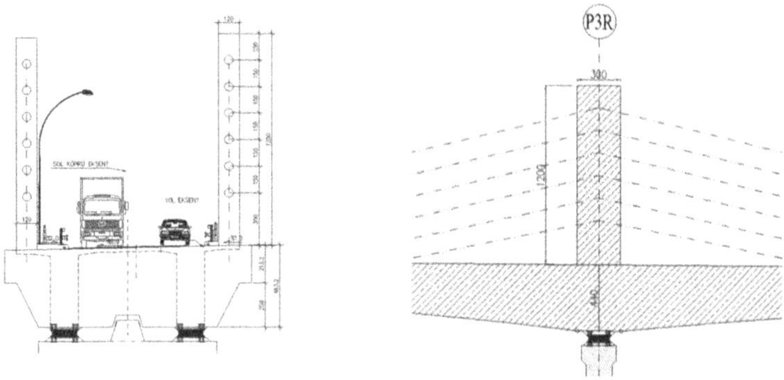

Fig. 4 Pylon views

Fig. 5 Saddles and saddle frame

In Dim Çayı Bridge, Freyssinet saddle system, which consists of a multi-tube saddle based on the use of Cohestrand stands, is used (Fig. 5). Cohestrand is a seven-wire strand coated with a fully bonded high-density polyethylene (HDPE) sheath. In multi-tube saddle, each strand of the cable is deviated individually by an individual sleeve, within the saddle tube and the bond of the Cohestrand sheath ensures the resistance to asymmetrical forces on either side of the saddle, by friction of the sheath on the individual sleeve.

This saddle system and the use of Cohestrands provide a complete and continuous corrosion protection ensuring over 100-year durability. It provides ease on application with mono-strand jacks and allows possible future individual strand replacements, if needed. Multi-tube saddle system also provides similar fatigue resistance with standard stay cable anchorages. In this project, 1860 MPa ultimate tensile strength capacity stay cables are used.

2.2 Seismicity

Dim Çayı Bridge is located in a moderate seismic area with the spectral acceleration of A0 = 0.22 g. The design spectrum for the 1000-year return period earthquake is generated per AASHTO Standard [4] using the following spectrum parameters: ground acceleration A0 = 0.22 g and soil class 2. The design spectrum used in the project is presented in Fig. 6.

The bearing arrangement is presented in Fig. 7. Twelve lead rubber bearings (LRBs) with Φ1000 mm diameter and 450 mm maximum displacement capacity and four pot bearings are used to satisfy service and seismic design criteria. At all piers, transversal shear keys are provided to transmit lateral forces to substructure (Fig. 8). The 100 m main span generates high axial dead loads on bearings. The production (diameter) and test capacity becomes an issue, if high axial load and high

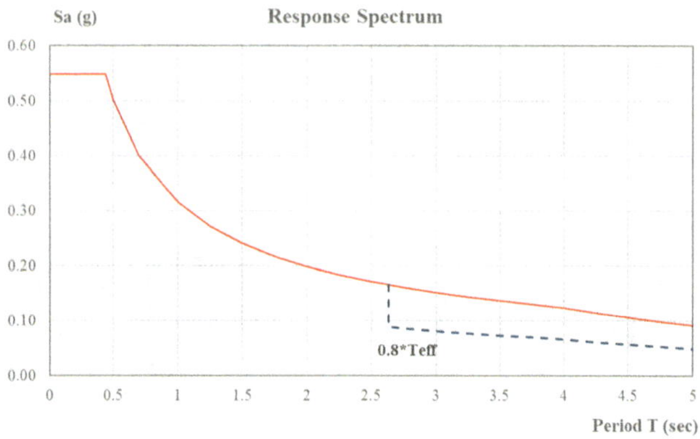

Fig. 6 Design spectrum

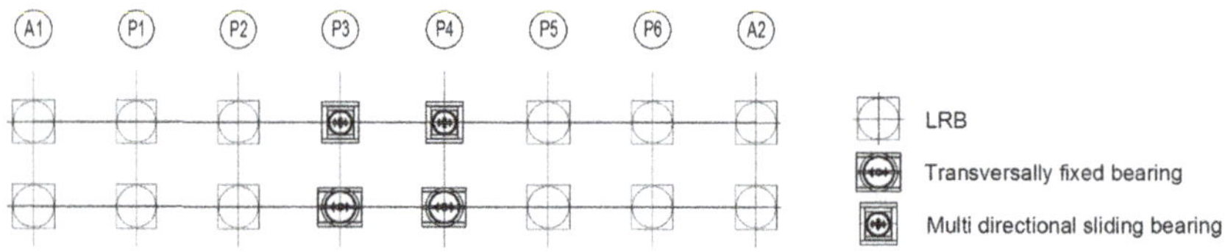

Fig. 7 Bearing arrangement

Fig. 8 Pier section

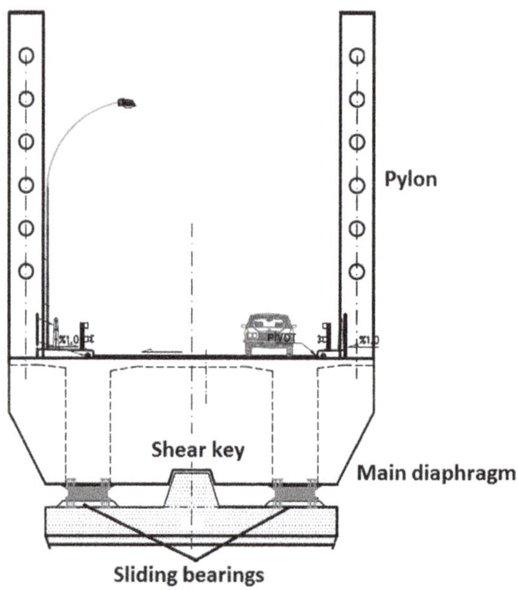

seismic displacement is required. In order to provide an economical solution, piers P3 and P4 are equipped with pot bearings, free to slide longitudinally and restrained transversally for service load conditions Remaining piers and abutments are equipped with lead rubber bearings (LRBs).

For global system analysis under seismic effects, effective stiffness and system damping provided by isolators are obtained from an iterative spectral analysis. The system damping converged to 27% after iteration. According to AASHTO Seismic Isolation Guide [5], the seismic spectrum is revised at periods greater than 0.8*Teff. The effective period is calculated using the converged stiffness and damping values. The multi-modal analysis is performed using Csibridge [6] and Midas Civil [7].

2.3 Construction Stage Analysis

Construction stages for classical extradosed bridges are important due to erection sequence of the extradosed cables. However, in Dim Çayı Bridge, total bridge length, deck shape and the presence of river under main span resulted to a complicated construction sequence. Contractor built the deck on limited amount of falsework with several casting stages, which is summarized in Fig. 9.

Construction started on one side of the bridge. After the piers were completed, first two spans were cast on falsework and tendons were stressed. In order to reduce the construction schedule, 3 out of 5 tendons on each rib were stressed on concrete blisters just before the construction joint, under the top slab. The remaining 2 out of 5 tendons were stressed on construction joint face, and continuity of these two tendons were provided using Freyssinet couplers.

For the next stage, formwork was moved to the adjacent location, third span and half of the main span was built in order not to block the whole river. During this stage, half of the riverbed in plan was filled with earth and the river was temporarily directed to other half, for falsework placement. The cantilever end of main span and midspan of the third spans are supported by steel temporary piers before the extradosed cables are installed (Fig. 10). This ensures controlled displacement, before the beam continuity is achieved. The same operation is followed for the remaining half, starting from the opposite abutment.

After the completion of deck construction, pylons were built. Then, cables were stressed one by one starting from right pylons. When the cable tensioning was completed, temporary supports were released and dissembled, temporary earth fill under temporary supports were removed, and the river was directed to the original bed.

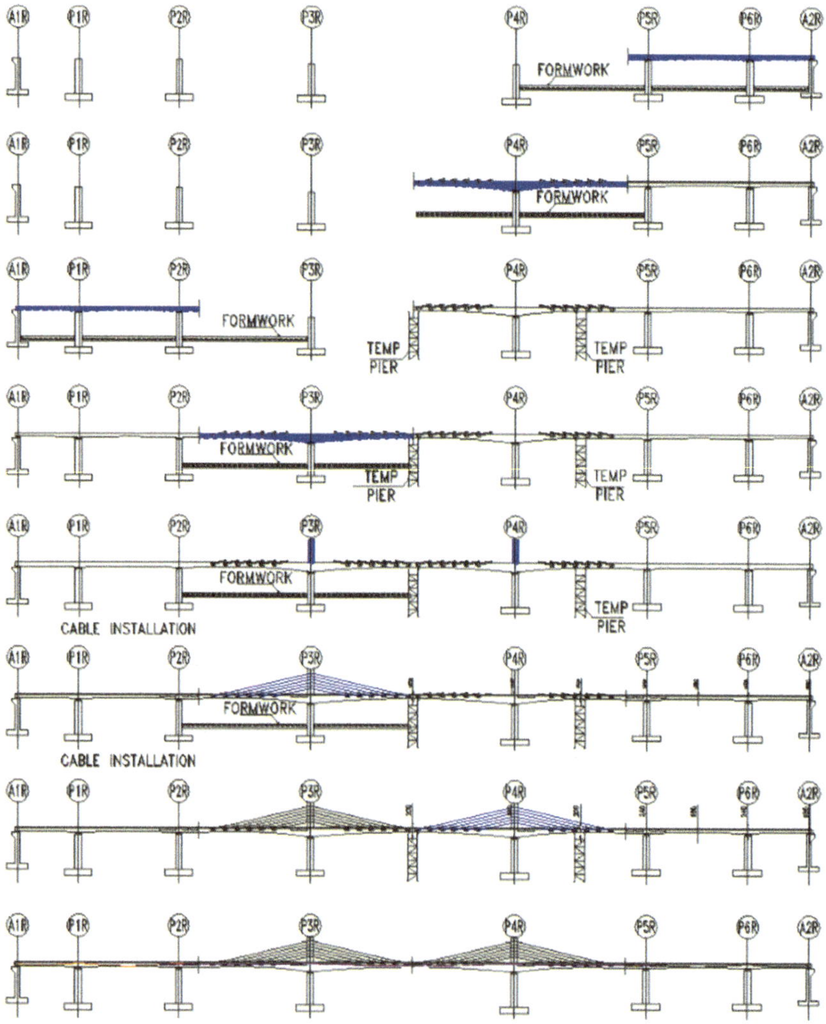

Fig. 9 Construction schedule

Stressing of stay cables was a delicate process. The cable installation was performed using the Freyssinet Isotension system, which adjusts the cable stress in each single strand tensioning. Isotension enables flexibility during installation while ensuring that forces within an extradosed cable are uniform. The deflection and cable forces were monitored throughout the construction, which confirmed theoretical estimations.

Fig. 10 Site photos during construction

3 Conclusions

In this paper, design and construction details of a river-crossing bridge Dim Çayı Extradosed Bridge is presented.

After the first successful application in Antalya, Çallı Extradosed Bridge, Dim Çayı has proven that extradosed bridges can be a competitive alternative not only in urban areas but also a river crossing in Turkey. The bridge is designed by Mega Mühendislik Müşavirlik A.Ş.. All extradosed cable works including design and procurement of saddles, expansion joints, bearings and post-tension equipment and installation is provided by Freysaş. The project will open to traffic in late 2018.

References

1. Mergamis KK (2008) Behaviour and design of extradosed bridges. University of Toronto, Canada
2. Picture reference (2001): https://structurae.net/photos/232383-kiso-gawa-bridge
3. Cable Stays (2002) Recommendations of French inter-ministerial commission on prestressing, SETRA, June 2002
4. AASHTO (2002) Standard Specifications for Highway Bridges, 17th edn. Washington, D.C., USA
5. AASHTO (2014) Guide Specifications for seismic isolation design 4th edn. Washington, D.C., USA

6. Computer and Structures, Inc., Csibridge Software Version 18.1.1
7. Midas Information Technology, Ltd, Midas Civil 2019

The Ethiopia Railway Viaducts: Steel Girder Launching and Permanent Bearing Design

Kutay Kutsal, Hatice Karayiğit, Cemal Noyan Özel, and Özgür Özkul

Abstract The Awash-Kambolcha-Hara Gebaya (AKH) Railway project is 389 km long single railway line over 54 bridges constructed in Ethiopia. The 5 m wide precast concrete composite deck rests on two built-up steel box girders side by side, supported on steel piers. The typical span is 46.4 m, formed by four 11.6 m segments using friction bolts. The construction of shallow pier and short bridges is performed using lifting cranes. For the construction of tall pier and long bridges, where the crane capacity is exceeded, incremental launching method (ILM) is used. Eight out of 51-bridges, ranging from 200 to 615 m continuous in length, are launched to provide construction ease and speed. The 11.6 m long steel box pieces are bolted at the assembly yard and launched over temporary bearings with an 18 m steel front nose. The longest bridge (B24) has 14 spans with 800 m horizontal radius, 2.42% slope and 45 m pier height. All launching equipment: back nose, pulling stick, guiding devices, temporary bearings and jack supports, are designed and manufactured locally in Turkey. The bridge design did not consider ILM method initially, so the authors worked with the already drilled existing web/flange splice bolt holes to connect the launching devices. In addition to the launching equipment, the permanent elastomeric bearings of all bridges are designed and manufactured in Turkey. The deck is supported on sliding elastomeric bearings longitudinally, to minimize seismic forces on the piers. Prestressed damping systems (PDS) are used in longitudinal direction to create a fixed point at one abutment in service state, while providing additional rigidity and damping during earthquake. The deck is fixed transversally using steel shear keys between the girders and the piers. The paper presents details about launching and bearing design of AKH railway bridges.

1 Introduction

Incremental launching method (ILM) is a bridge construction method that is widely used around the world. Many post-tensioned concrete, steel and composite bridges

K. Kutsal · H. Karayiğit · C. N. Özel (✉) · Ö. Özkul
Freysaş-Freyssinet Yapı Sistemleri AŞ, Istanbul, Turkey
e-mail: cemalnoyan.ozel@freysas.com.tr

P. Gülkan et al. (eds.), *Developments in International Bridge Engineering*, Springer Tracts on Transportation and Traffic 17, https://doi.org/10.1007/978-3-030-59169-4_14

have been constructed using ILM since 1960's. The method requires a construction area behind the abutment. The bridge sections are generally constructed behind the abutment then pushed into its final position. As a first phase of construction, temporary bearings and a casting beam are constructed. In the case of steel girder launching, casting beam is not required since the steel segments could be joined to each other on temporary bearings. Superstructure construction starts with front nose assembly. After the connection of the first segment with the front nose, each segment is launched over temporary bearings. The following segments are cast/bolted to the end of the last segment, and the sequence continues repeatedly until the bridge arrives into the final position. In order to achieve such a dynamic process, there are some equipment needed. The high strength steel strands coming from the hydraulic strand jacks located at abutment are connected to the pulling stick that is utilized to create a point in order to pull the bridge to the target location. Lateral movement during launching is blocked by guiding devices located at the piers and temporary bearings constructed in precast yard. Temporary bearings with PTFE pads are used in order to reduce friction between deck and the bearing surface. A front nose also attached to the tip of the leading segment in order to keep vertical displacement in limits. One of the most mechanized bridge construction method ILM is generally preferred since the bridges can be constructed over the areas having several constraints such as deep valleys, deep-water crossings, and steep slopes. Moreover, in the case of active traffic passage beneath the bridge, this method can be employed without any intervention to traffic.

For steel bridges, ILM has become an attractive method since it reduces the construction time, and provides safer assembly of the girders. Limited access for cranes at site, tall piers, and the need for high capacity cranes are other reasons to choose the ILM as a construction method for steel bridges.

In this paper, launching of the steel girders and the permanent bearings design in the scope of the AKH railway project are presented in detail.

2 AKH Railway Viaducts

2.1 General Overview

Ethiopian Railways Corporation (ERC) is the owner of the AKH railway project. ERC's aim is to improve the living conditions of the people and to develop the country by constructing the railway line. The project has a strategic importance in the development of the country as it connects the northern and eastern economic regions of Ethiopia. The owner's representative is Systra, and the contractor of the project is Yapı Merkezi. Tec4 is involved as the designer in the project. Freysaş is responsible for the ILM operation. Freysaş is also responsible for design and supply of bridge bearings and expansion joints.

Table 1 Details of launched bridged

Phase	Bridge	Slope (%)	Bridge geometry	Bridge length (m)	Launching direction
1st	B15	2.50	Straight	335.4 m	Upwards
1st	B24	2.42	curved, R = 800 m	613.8 m	Upwards
1st	B25	2.42	Curved, R = 800 m	289.0 m	Upwards
1st	B40	0.30	Straight	335.4 m	Upwards
1st	B43	0.06	Straight	196.2 m	Downwards
2nd	B46	2.50	Straight, L = 196 m	196.0 m	Upwards
2nd	B47	2.30	Curved, R = 1600 m	196.0 m	Upwards
2nd	B48	2.42	Curved, R = 1600 m	289.5 m	Upwards

The total length of the single railway line is 389 km, including over 54 bridges throughout the line. The project is planned to be built in two phases. The first phase is completed at the end of 2017. The second phase is initiated after the completion of the first phase, and the line is planned to be completely in use at the end of 2019. The bridges having shallow piers and short length was constructed using lifting cranes. However, when crane lifting capacity is exceeded, ILM is employed in order to provide speed and ease. The number of the bridges launched in the first phase are five, and three bridges are to be launched in the scope of the second phase. These bridges are ranging from 200 to 615 m continuous in length, and the details are given in Table 1.

2.2 *Details of Launched Bridges*

The superstructure of the bridges is designed as 5 m wide composite deck consisting of precast concrete slab connected to the two built up steel box girders with shear studs (Fig. 1).

The start and end spans of the bridges are 28.50 m, while the typical span is 46.4 m. Typically 46.4 m span is formed by four 11.6 m segments connected to each other with (10.9) quality pretensioned bolts. B24 is the longest bridge launched in the project, having 14 spans with 800 m horizontal radius, 2.42% slope and 45 m pier height (Fig. 2).

The required jack capacity is determined considering the longest bridge (B24). 18 m front nose is attached to the front of the steel box girders, which are already assembled in casting yard. After the steel box girders assembled in casting yard, completed part is slided over temporary bearings (Fig. 3).

In order to keep the vertical deflection within acceptable limits, first segment of each bridge is launched without precast concrete slab. By doing so, smaller size of front jacks are used, located at the tip of the front nose in order to compensate the level difference when nose reaches the piers.

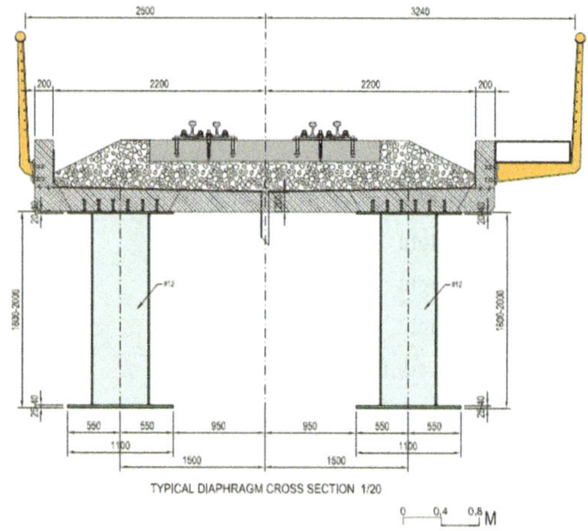

Fig. 1 Typical cross section of the deck

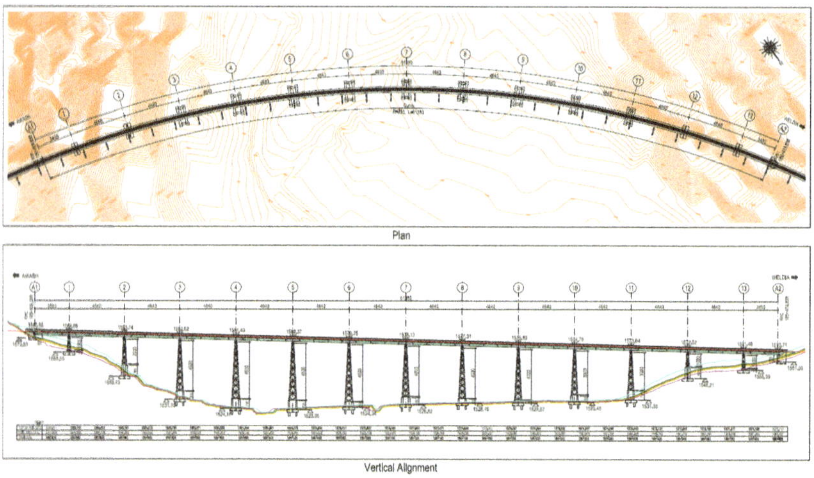

Fig. 2 Plan view and elevation of the B24 bridge

All launching equipment: back nose, pulling stick, guiding devices, temporary bearings and jack supports, are designed and manufactured in Turkey.

During the construction, the steel girders are supported by two temporary bearings, which enabled the sliding of the deck with a low friction coefficient on each pier. Launching bearing is formed by a steel pedestal, elastomeric pads and cover plate. Temporary bearings are bolted to the plate on pier cap. All transversal forces due to launching and wind is assumed to be resisted by guiding devices. Friction forces due to vertical load are resisted by launching bearings, shear keys and bolted connection (Fig. 4).

Fig. 3 The bridge on temporary supports

Fig. 4 Guiding device, temporary bearings, lifting jack

The launching jack is installed on the front surface of the jack support plate using 4 M30 bolts (8.8). On the concrete side (back surface), the plate has 510 mm long seven steel studs. This plate is quite important to allow for the transfer of loads coming from the jack to the abutment (Fig. 5).

The pulling stick, which pulls the bridge to determined location by means of the strands coming from the abutment, is attached to the 2 m long back nose with welded connection. The purpose of the back nose, which has the same geometric properties with typical bridge girders, is to bring the bridge to final location after last launching process (Fig. 6).

As the bridges is initially not considered to be constructed using ILM, already drilled existing web/flange splice bolt holes are used in order to connect the back

Fig. 5 Jack support and strand jack

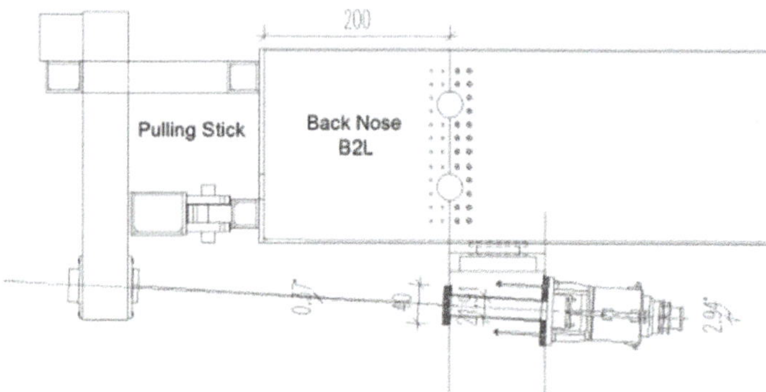

Fig. 6 Back nose-extension detail

nose and the steel girders. Back nose is connected to the steel girders with M36 (10.9) bolts at flange, and M30 (10.9) bolts at web. The bolts used during launching are pretensioned to 40% of capacity in order to be utilized repeatedly.

Bolted splice connection between back nose/steel girder and welded connection between pulling stick/back nose connection are designed to resist the maximum pulling force, corresponding to the last launching effort of the bridges (Fig. 7). In order to determine the effects on the pulling stick under the launching forces, the members are analyzed in SAP2000 [4]. The members of the pulling sticks are dimensioned according to AISC 360-10 [3], after the static linear analysis has been performed.

There are some challenges faced during design stage in order to provide the connection between back nose and steel girder. The bolted connection is designed as the same with typical girder splice. However, it had to be modified for the last launching. The reason of this modification is that the last segment has no splice holes, as it is located at the end of the bridge. There are also more shear studs on top flange

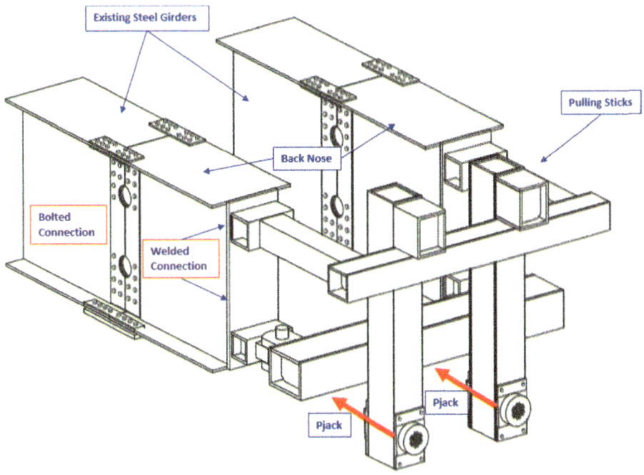

Fig. 7 Pulling stick-back nose and steel girder connection

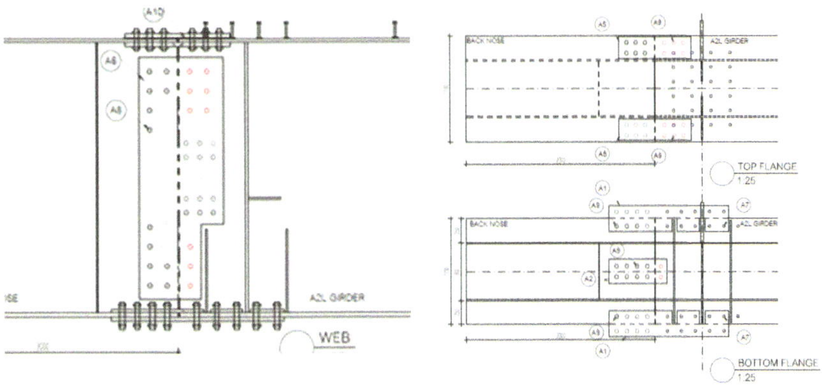

Fig. 8 Modified back nose-steel girder connection

than the typical segments, and additional vertical stiffeners are located at the webs. In order to provide the connection, additional bolt holes are drilled at flanges and webs of the girders. In addition, existing bolt holes for prestressed damping systems (PDS) connection are also used. Bottom flange plates are designed in such a way that they do not intersect with the vertical stiffeners (Fig. 8).

2.3 *Permanent Bearing Design*

The steel composite deck is supported on elastomeric bearings in longitudinal direction in order to keep the seismic forces on the piers minimum. In the transverse

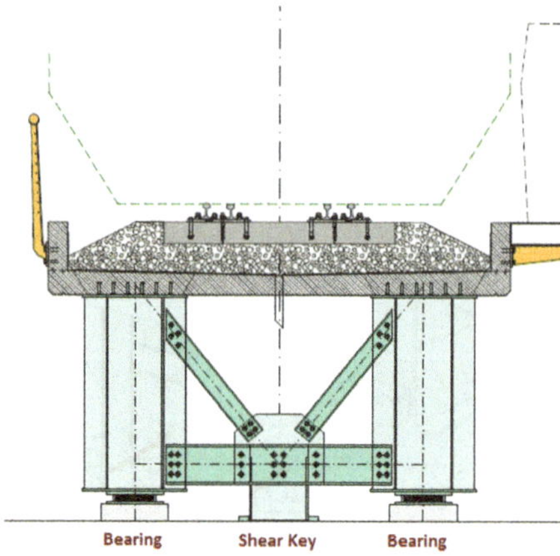

Fig. 9 Transversal section of the deck at diaphragm

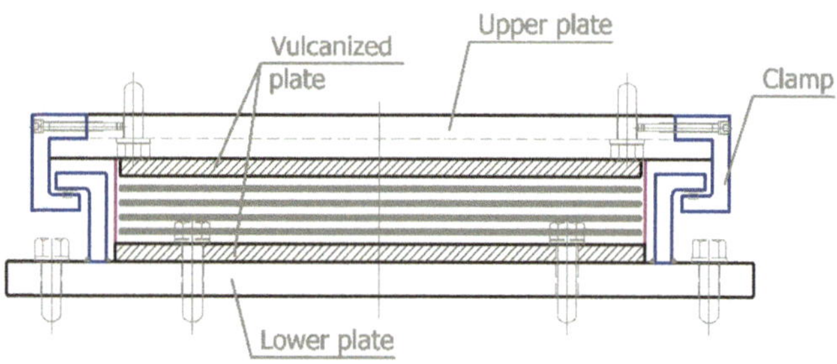

Fig. 10 Anti-uplift elastomeric bearing

direction, the steel shear keys attached to the bottom flange of the deck are used in order to restrain the bridge (Fig. 9).

The elastomeric bearing types used are, free in both directions, fixed in longitudinal direction and sliding in the longitudinal direction. 14 different types are determined based on design parameters: design load, displacement capacity and rotation capacity. The bearing design is performed per EN 1337-3 [1] and EN 1998-2 [2] is taken into account for the seismic condition design checks. The 346 of the total 498 elastomeric bearings are designed to withstand uplift force (Fig. 10).

When the span number is greater than four, the bridge system is restrained at one abutment using Prestressed Damper System (PDS) under service condition. Use of the PDS system results in a small reaction on the piers by providing additional stiffness and damping under seismic condition. 57 prestressed dampers are used in the project.

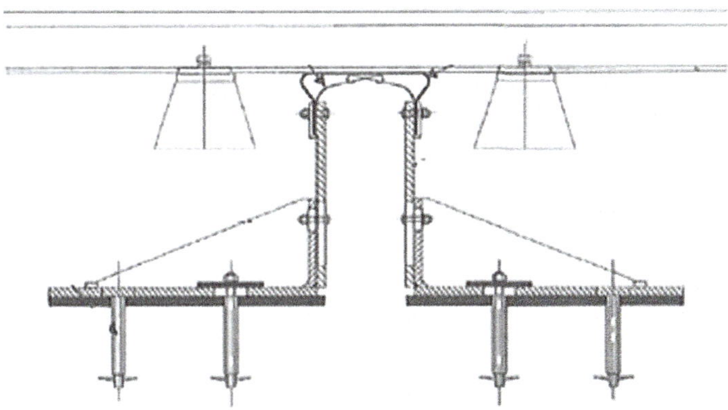

Fig. 11 Retaining ballast section

Fig. 12 Application of the retaining ballast

Three types of expansion joints are used based on displacement capacity. The expansion joint displacement capacity (from smaller to larger) are TW joint, Sub-ballast, and Retaining ballast, and the displacement values change between 20 to 602 mm (Figs. 11 and 12).

3 Conclusion

This paper gives general information about the AKH Railway Project and presents the details of the steel girder launching. AKH railway project is an important project for Ethiopia as it connects the northern and eastern economic regions of the country. The challenges faced during the design stage in order to apply ILM are explained in detail. General information about the launching equipment, which is designed and manufactured in Turkey, are provided. Moreover, the details of the permanent bearings, which is also designed and manufactured locally in Turkey, are presented.

In the scope of the project, eight bridges are constructed using ILM. Since it provides ease and speed, this method proved to be the best solution for the construction of these bridges. The project is planned in two phases. The first phase is completed successfully, and the second phase is underway. The railway line is planned to be opened by the end of 2019.

References

1. Eurocode EN 1337-3 (2005) Structural bearings – Part 3: Elastomeric bearing, Brussels
2. Eurocode EN 1998-2 (2005) Design of structures for earthquake resistance – Part 2: Bridges, Brussels
3. AISC (2010) ANSI/AISC 360-10, Specification for structural steel buildings. American Institute of Steel Construction Inc., Chicago
4. SAP (2000) Structural analysis program. Computers and Structures Inc., Berkeley

Incremental Launching by Lag-Casting: İhsaniye Viaduct

Cemal Noyan Özel, Özgür Özkul, and Hatice Karayiğit

Abstract Incremental Launching Method (ILM) is a bridge construction technique that has become an efficient alternative in Turkey in recent years. In the first phase of The Northern Marmara Motorway Project (KMO1), three viaducts were constructed using ILM. The KMO1 is followed by the second phase: KMO2. İhsaniye Viaduct is part of KMO2 on the European Side of Istanbul, located south of the Third Istanbul Airport (IGA), and constructed using the ILM. Several advantages are offered by the ILM technique, including quantity saving, enhancing the safety during construction and introducing innovative seismic design approach. This paper focuses on the deck and pier design of the İhsaniye Viaduct. The deck is constructed by lag casting of segments. First, only the bottom slab and the webs are cast. After reaching the launching strength, the U-section is pushed out-of-the formwork, where the top slab is cast simultaneously with the next segment's U-section in the second step. This technique provided a fast launching track, optimizing the time required for strength and workforce. Detailed analysis of the construction staging, post-tensioning and rebar detailing are exhaustively studied. The pier design is governed by the seismic actions. Innovative double-wall pier shape was used to increase its flexibility in the transversal direction while providing energy dissipation through the creation of plastic hinges at double walls. Longitudinally, the deck is fixed on several piers and fluid viscous dampers are placed at the abutments to reduce the seismic displacement. The viaduct is two decks side by side, 21.5 m wide, 856 and 867 m in length with 66 m typical spans and an 80 m maximum span. The 80 m span is launched using a temporary steel pier.

1 Introduction

Technological developments have brought many advantages and innovations to different construction types as well as bridge construction methods. Incremental Launching Method (ILM) is one of the highly mechanized and efficient way to build

C. N. Özel (✉) · Ö. Özkul · H. Karayiğit
Freysaş-Freyssinet Yapı Sistemleri AŞ, Istanbul, Turkey
e-mail: cemalnoyan.ozel@freysas.com.tr

P. Gülkan et al. (eds.), *Developments in International Bridge Engineering*, Springer Tracts on Transportation and Traffic 17, https://doi.org/10.1007/978-3-030-59169-4_15

concrete, steel or composite bridge decks. ILM has been used all around the world since 1960s in different ways and has become an economic and realistic alternative in Turkey with successful examples in recent years. In the first phase of the Northern Marmara Motorway Project (KMO1), three viaducts are constructed using ILM with deck lengths vary between 282 to 642 m.

With ILM, 40–70 m spans with up to 1200 m total continuous deck lengths can be built. In addition, it is possible to increase the maximum span length using temporary piers during the launching operation. This method requires a casting yard, generally behind the suitable abutment where the deck is cast segment-by-segment. Casting yard length depends on typical span length, construction schedule and other project based restrictions. Construction sequence starts on casting yard, deck is formed on mechanized formwork system, post-tensioning is applied then deck is pulled/pushed on temporary-sliding bearings using hydraulic jacks. This cycle of launching goes on for each successive span over piers creating a continuous structure. When pushed in location, service post-tension tendons are stressed and temporary bearings are replaced with the permanent bearings. Launching operation for a concrete deck can take 5 m per hour depending on deck weight and slope of the bridge.

ILM brings many advantages, such as safety, quality, material quantity saving and reduction in construction time. In addition, it is also possible to further reduce material quantities of substructure (piers and foundations) in high seismic zones with changing bearing layout and pier shapes efficiently.

Ihsaniye Viaduct which is an important part of second phase of Northern Marmara Motorway (KMO2) is constructed using ILM. In this project to reduce construction cycle time, an alternative deck lag casting is used. Details of deck construction and innovative seismic design of the piers are presented in this paper.

2 Ihsaniye Viaduct

2.1 General Information

Northern Marmara Highway (425 km in total length) is the new motorway to bypass Istanbul from Silivri (in European side) to Akyazı (in Asian Side) to the north and makes a connection to the Third Istanbul Airport. First phase (KMO1) of motorway (about 95 km in length) which includes Yavuz Sultan Selim Bridge over Bosphorus is completed and opened to traffic in 2016. Second phase of the project (KMO2) has started in 2017 in European and Asian sides. Marmara Sea and Ihsaniye project location is presented in below Fig. 1.

Ihsaniye Viaduct is part of KMO2 in European side of Istanbul and just south of the Third Istanbul Airport. In the beginning of the project, the viaduct was designed as conventional simply supported I-girder deck with 40 m spans. However, the viaduct route crosses a double layered intersection of another motorway, and one pier couple axis overlap with this intersection. Therefore, contractor evaluated different options

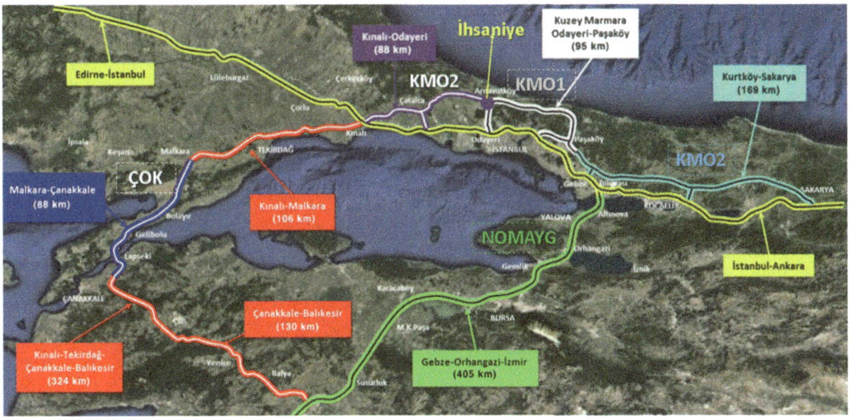

Fig. 1 General view of existing and planned motorways in Marmara Region [1]

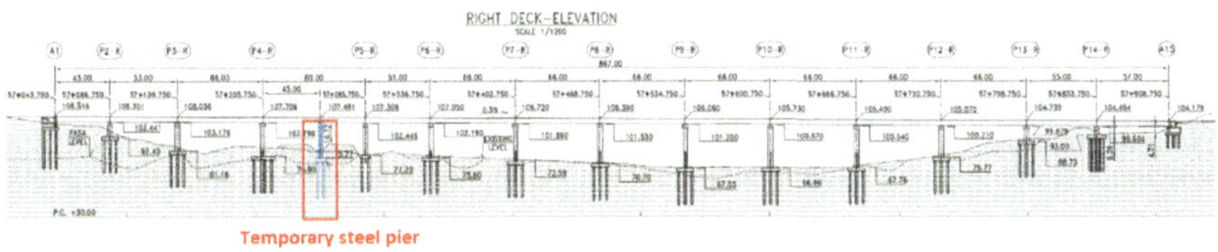

Fig. 2 Elevation view of right deck

in order to solve this overlapping problem and time constraints. After evaluations, ILM method is chosen due to its longer span, construction time and material quantity advantages.

2.2 *Bridge Details*

The viaduct is 856 and 867 m long, two parallel decks with five traffic lanes. Typical span length is 66 m. At the intersection location, a temporary steel pier is used during launching operation (Fig. 2). After construction is completed, the temporary pier will be removed and total span length will be 80 m in service state. Number of piers and foundations is reduced, as compared to the precast I-girder option, which provided a considerable material saving. The viaducts are straight in plan (Fig. 3).

The deck is designed as a post-tensioned concrete single cell box girder with 4.00 m constant depth and 21.5 m width (Fig. 4). Additionally, in deck top slab, transversal posttensioning is used to reduce the top slab thickness. Transversal post-tensioning tendons (4B15) are used and stressed at one end by staggering.

The pier consists of two different sections: double wall section at the top and I-shaped rigid shaft at the bottom. The height of the double wall, which has a

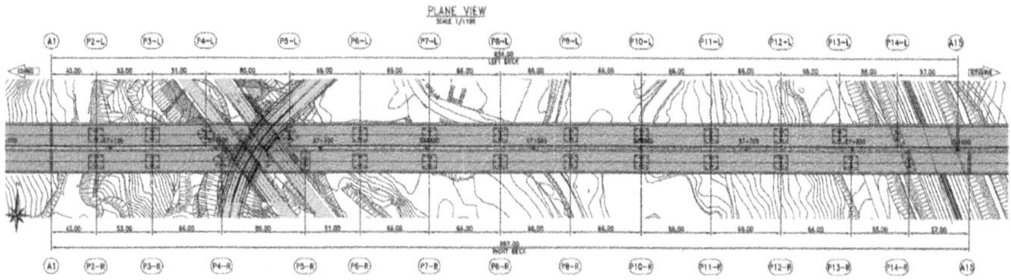

Fig. 3 Plan view of viaducts

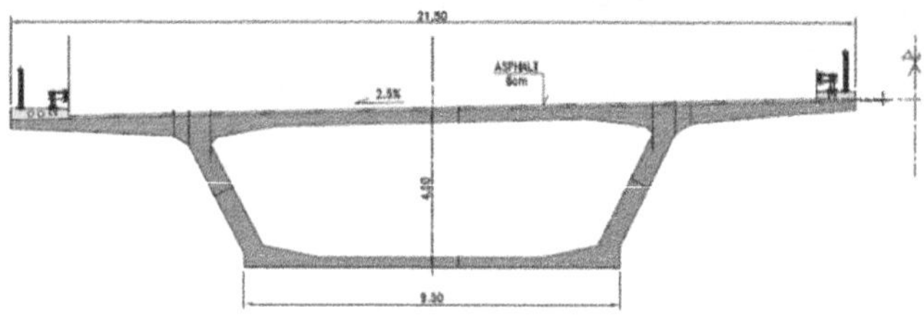

Fig. 4 Deck section at the mid-span

Fig. 5 Details of the pier section

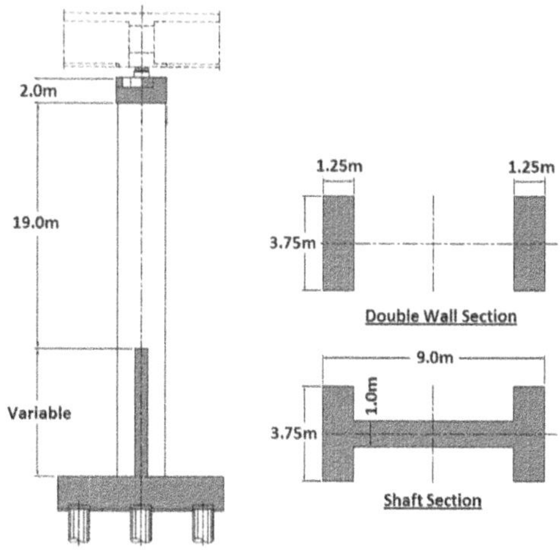

3.75 * 1.25 m rectangular cross section, is limited to 19 m height. In the case where piers are taller than 19 m, I-shaped shaft section is used as shown in Fig. 5.

2.3 *Seismic Design*

Istanbul is located close to a high seismic Northern Anatolia Fault Zone. Due to location and importance of the project, seismic design of the substructures has become an important part of Ihsaniye Viaduct. The main purpose of seismic design is to safeguard to major failures on structure and loss of life. The 1000 years return period earthquake is used for permanent structure design, and the design spectrum (Fig. 6) is generated according to AASHTO LRFD [2] with site-based spectral acceleration parameters. Under this level of earthquake, structures can respond in the inelastic range with a ductility ratio not exceeding the allowable ductility demand R.

For Ihsaniye Viaduct, an alternative and innovative seismic design strategy is used in order to provide safety and economy, which was also used in the first phase of Northern Marmara Motorway (KMO1). For these purposes, pier shapes, bearing types and arrangements are revised from the original design. Seismic design strategy aims firstly limiting the longitudinal displacements with fixed points (fixed piers by pot bearings). Other piers are isolated from the deck with sliding pot bearings. This isolation provides the system extra flexibility. Additionally, in order to reduce the seismic loads (forces and displacements), fluid viscous dampers are provided at each abutments. These dampers provide additional damping and extra stiffness during an earthquake event, to eliminate any irreversible damage to the structure. In the transverse direction, seismic shear keys are provided at top of the taller piers and abutments, and the vertical sliding surfaces are used at the contact surface of the pier cap and shear key to transfer the horizontal load and allow the deck displacement in the longitudinal direction. The shorter piers are considered as transversally free due to their higher rigidity. It is assumed at certain locations of piers (top and bottom) plastic hinges will be formed in transverse direction, which can be repaired in a short term. The bearing arrangement of the system and the vertical sliding surfaces are presented in Figs. 7 and 8.

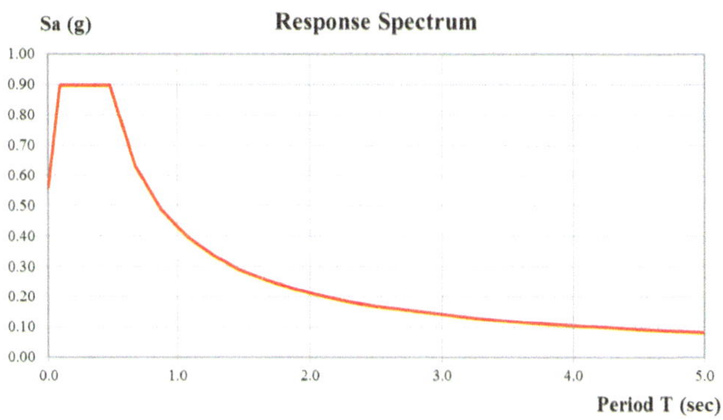

Fig. 6 Design spectrum

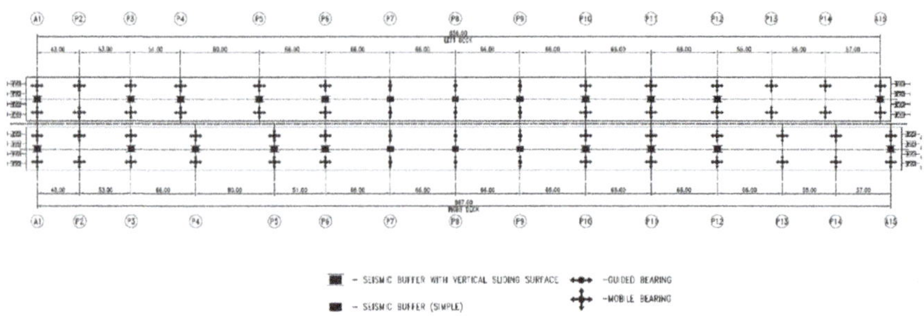

Fig. 7 Bearing arrangement

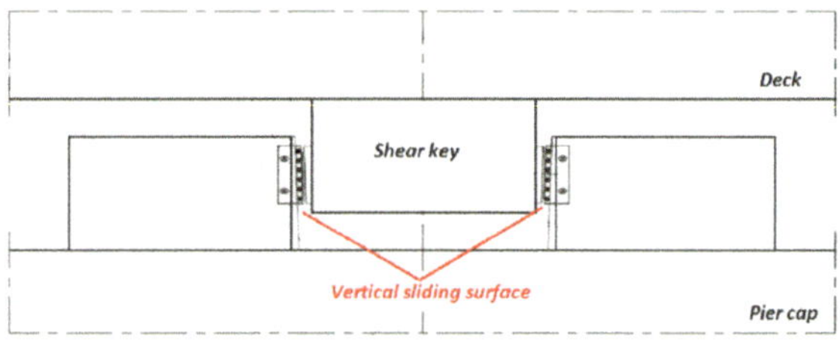

Fig. 8 Details of the vertical sliding surface

To provide a uniform stiffness and uniform load distribution in the transverse direction, top part (19 m) is formed as less stiff double wall portal frame under solid pier cap. The remaining bottom part (if pier is taller than 19 m) is made of stiffer I-Shaped beam. Top and bottom zone of the double walls are reinforced to allow plastic hinges.

2.4 Construction and Launching

To reduce the total construction time and finish the project in given time frame, typical construction cycle of ILM is needed to be altered. For this purpose, an alternative is proposed: Incremental launching by Lag-casting. In this case, casting yard needs to be a bit longer than the conventional full casting. Casting yard is divided into two parts with two different formwork system: top casting area and U-casting area. In U-casting area, which is designed as 33 m long, only the bottom slab and the webs are cast as U shape. After concrete is reached a certain strength, the unstressed U part is launched to top casting area (which is 33 m long) where the top slab part is cast. By doing so, the production is divided into two and segments are built in operational sequence staggered in both time and space. Availability of each activity time is increased and critical situations are reduced [3].

Fig. 9 Precast yard

Fig. 10 Typical construction sequence

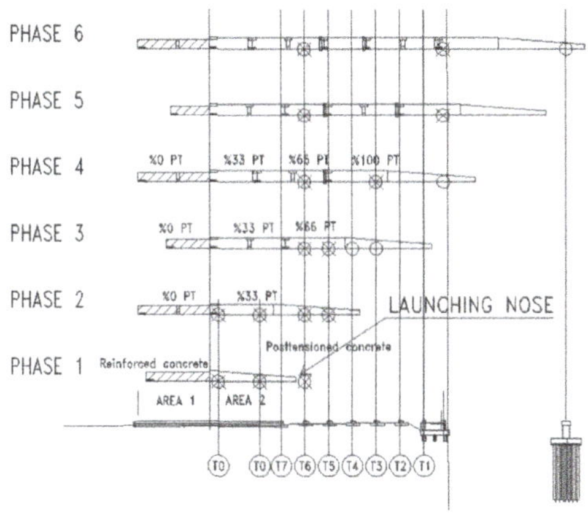

The total length of the precast yard including casting beams and temporary supports, which are used for launching nose assembly and stressing the remaining tendons, is about 135 m (Fig. 9).

Typical construction sequence is presented in Fig. 10. In U-casting zone (Area 1) bottom slab and webs are cast, but are not stressed. Then, the reinforced concrete U-section is pulled to the top casting area (Area 2). After top slab is cast, %33 of final launching posttensioning is applied. The stressing of post-tension tendons, which is needed for launching, is completed in three phases to %100, before the deck reaches the abutment. With lag-casting application in Ihsaniye, construction speed is increased. A full cycle is completed in 4 days after learning curve.

In order to avoid high shrinkage differential between the U-section and the box section, time between two castings is recommended not to exceed 3 days in AASHTO [2]. When the top slab concrete reaches the required stressing strength, launching tendons are jacked. Launching tendons are located at top and bottom slab of the deck. Launching tendons are straight in profile. Concentric tendons generate only

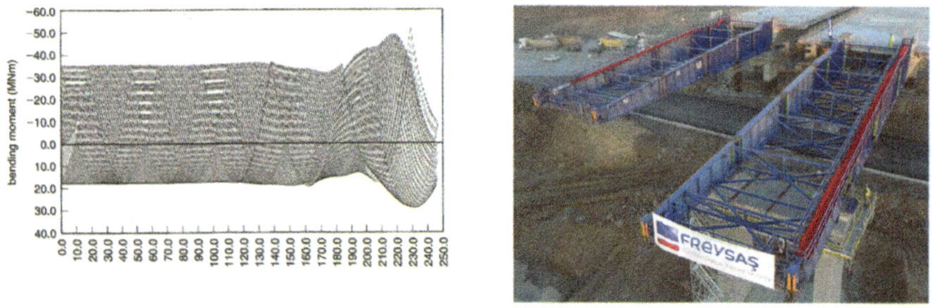

Fig. 11 Typical launching moment envelope and launching nose

axial compressive force, but not bending moment. During the launching operation, each section of the deck is subjected to both tensile and compressive stresses. Typical bending moment diagram of deck during and the launching nose is shown in Fig. 11. 13c15, 15c15, 19c15 type Freyssinet anchorages are used for launching tendons in Ihsaniye Viaduct.

From Fig. 11, cantilever end moment diagram is larger than the moment in typical spans. To limit this extra cantilever moment, a special equipment "launching nose" is used. Launching nose is generally made by steel sections, which is lighter and less stiff than concrete. It reduces the cantilever moment and provides a more economical solution. In Ihsaniye Viaduct, the steel launching nose is composed of two parallel I-girders with varying height. The nose length is 40 m, and is connected to the concrete deck using high strength Freyssibars.

Deck is launched to upwards with 0.5% longitudinal slope using two hydraulic jacks. To reduce friction at bearings during launching, special sliding bearing with PTFE surface is used. Transversal movement of deck is restrained using guiding devices on piers.

Ihsaniye Viaduct passes over two active motorways. One is under the first span at the launching abutment side. The other underpass is under the 80 m span. During the launching operation, these two motorway underpasses are not stopped or closed, and continued their operation. A slender steel temporary pier with 15 m height is installed around the midspan of the 80 m span during launching. Temporary piers are shown in Figs. 12 and 13. Deflections on piers, temporary piers and decks are monitored during all construction steps.

After launching is completed and the deck reaches its final position, eccentric post tensioning tendons are stressed, which are provided in addition to the existing concentric launching tendons. These external service tendons are placed inside the deck and housed in HDPE ducts, creating a harped at two points profile. Tendons are deviated by concrete deviators and diaphragms inside the deck. 19c15, 25c15 and 31c15 Freyssinet anchorages are used for external tendons.

Fig. 12 Temporary piers

Fig. 13 Eighty-meter span and temporary pier

3 Conclusions

This paper presents details of construction and design of Ihsaniye Viaduct, which is an important part of Northern Marmara Highway. The Incremental Launching Method modified using lag casting is presented, which provided many advantages to construction schedule. Moreover, innovative seismic design strategy and calculation principles are used for pier design.

The main contractor of the project is Avrupa Otoyolu Yatırım ve İşletme AŞ, designer of the viaduct is Nergiz Group Proje AŞ. Freysaş is responsible for deck construction, supply and execution of post-tensioning tendon and anchorages, temporary/permanent bearings, expansion joints, FVD's and temporary piers procurement and installation.

Ihsaniye Viaduct is another successful example of ILM bridges build in Turkey in recent years. These examples show that ILM is an economic and effective alternative for future projects.

References

1. 1915canakkale.com.https://www.1915canakkale.com/assets/GuzergahImg/guzergah-tr.jpg Accessed 29 Aug 2018
2. AASHTO LRFD (2010) Design specification, Washington, DC
3. Rosignoli M (1998) Launched bridges: prestressed concrete bridges built on the ground and launched into their final position. ASCE Press, Reston

Effect of Skew Angle on the Rotation of Exterior Girders During Construction

Faress Hraib, Li Hui, Miguel Vicente, and Riyadh Hindi

Abstract Bridge designers tend to extend the deck slab width beyond the exterior girders, over a distance called the overhang, to increase the width of the bridge without adding extra girders. Moreover, the screed machine used in finishing the surface of the deck slab usually rests on the edges of the overhang. The machine weight combined with the weight of the fresh concrete leads to torsional moments in the exterior girder. Hence, excessive rotations in the exterior girder arise and they lead to issues such as lose of concrete cover, instability of the superstructure and non-uniform deck slab thickness. Many Departments of Transportation are facing this problem and there are no specific guidelines to compute this rotation. Contractors usually use temporary bracing systems such as timber blocks and diagonal or transverse ties to reduce the rotation. However, these methods were found to be not very effective. The bridge geometrical parameters play a significant role in this phenomenon, and one of these parameters is the skew angle. Skewed bridges are popular because of the geometrical conditions imposed by roads alignment and the geographical barriers to cross, which leads to a non-perpendicular crossing. In these cases, skewed bridges are more economical than an equivalent straight bridge. This study focuses on the effect of skew angles on the rotation in steel girder bridges. The purpose of this study is to examine different skew angles using Finite Element Analysis (FEA) of bridges subjected to construction loads. SAP2000 is used to develop the FE models of these bridges. The findings of this research will provide designers and contractors a better understanding of the effect of skew angles on the rotation of exterior girders

---

F. Hraib (✉) · L. Hui · R. Hindi
Parks College of Engineering, Aviation and Technology, Saint Louis University, St. Louis, USA
e-mail: faress.hraib@slu.edu

L. Hui
e-mail: li.hui@slu.edu

R. Hindi
e-mail: riyadh.hindi@slu.edu

M. Vicente
School of Civil Engineering, University of Burgos, Burgos, Spain
e-mail: mvicente@ubu.es

P. Gülkan et al. (eds.), *Developments in International Bridge Engineering*, Springer Tracts on Transportation and Traffic 17, https://doi.org/10.1007/978-3-030-59169-4_16

and opens the way for future research to develop methods to reduce the rotation to satisfactory limits.

1 Introduction

During the construction of composite bridges, screed machine is used in finishing the concrete surface of the deck slab. Combined with the self-weight of the fresh concrete, these two loads create rotational moments in the exterior girders when they act on the overhang [1]. The overhang is that part of the deck slab that extends beyond the exterior girders in the transverse direction to increase the slab width without adding more girders. The overhang formwork is usually supported by steel brackets that rest on the exterior girders as shown in Fig. 1 and it acts as a cantilever supported by the exterior girder.

Therefore, when the rail of the screed machine is placed on the edge of the overhang it creates those rotation moments that lead to excessive rotation of the exterior girder [2]. This rotation can lead to stability issues, future maintenance issues, reduction in the deck slab cover and many other concerns [3]. Timber blocks, transverse ties and diagonal ties are very common examples of traditional rotation prevention systems that contractors tend to use. Recent studies have proven that these systems fail to resist the rotation and reduce it to the acceptable limits [4]. Moreover, Ashiquzzaman et al. showed that the ratio between the lateral bracing system members (B) and the depth of the girder (D) is a good indicator to limit the rotation of the exterior girders [5]. Also, it has been proven that adding a temporary bracing systems at the exterior bays of the superstructure can reduce the rotation of the exterior girders significantly [4].

On the other hand, skewed bridges are very popular because of the geometrical conditions imposed by roads alignment and the geographical barriers to cross, which leads to a non-perpendicular crossing. In these cases, skewed bridges are more

Fig. 1 Overhang formwork

economical solution than an equivalent straight bridge. However, bridge skewness will lead to complex issues in the structural behavior of the bridge during and after the construction. This research focuses on investigating the effect of the skew angle on the rotation of the exterior girders during bridge construction. Also, it focuses on the effectiveness of the temporary bracing systems that are used to reduce the rotation. Finite Element (FE) models of steel girder bridges are developed in this study to examine the effect of skew angles, and the different temporary bracing systems on the rotation of exterior girders during construction.

2 Finite Element Modeling

Using SAP2000 several Finite Element models are developed to simulate the behavior of bridges subjected to construction loads. The bridge superstructure is modeled using a combination of shell and beam elements. Girders' and diaphragms' flanges and webs are modeled using 4 nodes thick shell elements with the corresponding thickness assigned to each member. Steel brackets, cross frames, pipes, ties and screed rail are simulated using 2 nodes beam elements. Pin and roller boundary conditions are applied at the nodes of the bottom of the girders that correspond with the location of end abutments or intermediate piers as shown in Fig. 2.

Concentrated loads applied directly to the nodes are used to simulate the fresh concrete and the screed machine loads. The screed machine wheels are placed on the screed rail at the edge of the overhang on each side of the bridge. Four wheels on each side, each is 3.34 kN (750 lbs) and they are longitudinally spaced at 0.61 m (2 ft), 1.22 m (4 ft) and 0.61 m (2 ft). The fresh concrete load is distributed between girders based on the concept of tributary area. As the screed machine moves along the span

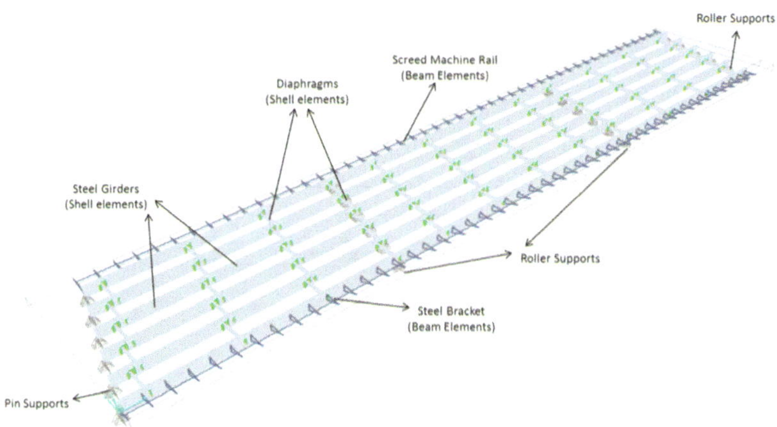

Fig. 2 Finite element modeling

of the bridge to level the concrete surface, all FE models are loaded following the same notion. The traveling step is taken equal to 1.22 m (4 ft). Hence, with each step a loading case is created to simulate the screed machine movement and the concrete loads follow the screed machine until the deck slab is fully loaded. It also should be noted that the screed machine is always perpendicular to the girders regardless of the skew angle of the bridge. The FE model used in this study was verified with a field monitored bridge during the construction, and the FE model was able to predict the rotation recorded in the field with an acceptable error of 3.2%.

3 Bridge Description and Skew Angle Variation

A skewed steel girder bridge located in Ford County, Illinois, USA was chosen to be analyzed using SAP2000, and then examine the effect of changing its skew angle. The bridge has a skew angle of 14° and consists of six W27 × 161 that extends on three spans of 17.53 m (57.5 ft), 22.71 m (74.5 ft) and 17.53 m (57.5 ft). The girders are equally spaced at 2 m (6 ft and 8 in.), and the overhang width is 0.9 m (35 in.). The diaphragms are C12 × 25 and they have a maximum spacing of 5.37 m (17.625 ft) as shown in Fig. 3.

As the bridge skewness is the focus of this study, the original skew angle of this bridge (14°) was changed to (0°), (28°) and (42°), while keeping all other geometrical parameters of the bridge constant. The bridge diaphragms were also held in a staggered formation to isolate the effect of the skew angle. In each bridge model with different skew angle, the screed machine was moved from the beginning to the end of the bridge along with the corresponding concrete load. At each loading step the rotation values along the exterior girders (G1 and G6) were recorded at three locations: Top, Middle and Bottom of the web of the girders.

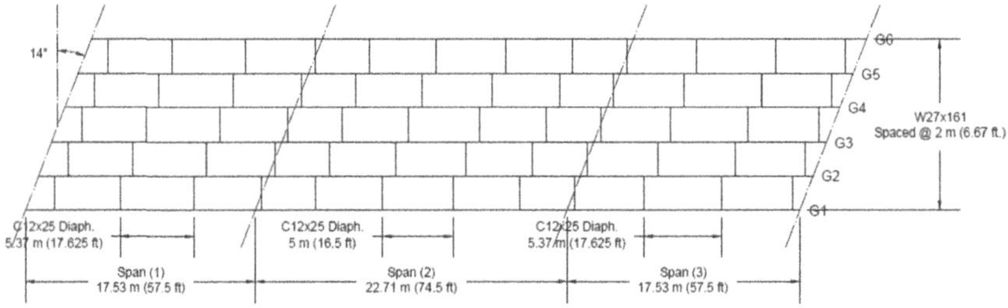

Fig. 3 Ford county bridge description

4 Results and Discussion

The results of the FE modeling of Ford County Bridge are discussed in detail in this section, highlighting the following: the maximum rotation location in the longitudinal direction of the bridge, the effect of different skew angles and the effectiveness of B/D ratio application with temporary bracing systems.

4.1 Location of Maximum Rotation

The maximum rotation was observed in both exterior girders at approximately the middle distance between the two intermediate diaphragms in spans 1 and 3, as shown in Fig. 4. However, because the screed machine is perpendicular to the girders, the maximum rotation occurs in exterior girder G1 before it does in exterior girder G6. In fact, when the rotation value in exterior girder G1 reached its maximum of 0.718°, the rotation in exterior girder G6 was only 0.459°.

The rotation values along the web of exterior girder G1, and exterior girder G6 are shown in Fig. 5 and 6, respectively. These values are recorded at the top, the middle and the bottom of the web of the girder for the loading case that produces the maximum rotation in each exterior girder. For exterior girder G1, case (40) produced the maximum rotation value of 0.718°, while it was 0.620° for exterior girder G6 in case (43).

The results show the same pattern between the two exterior girders, where the rotation value maximizes between the intermediate diaphragms and reaches values close to zero at the diaphragms locations. However, it also proves that because of the bridge skewness there is a difference between the rotations of exterior girders.

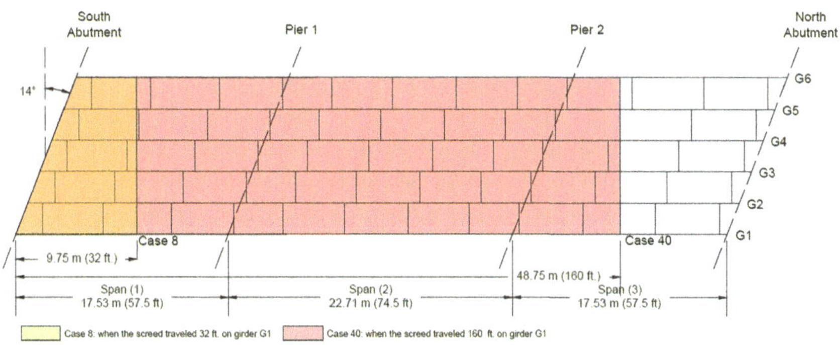

Fig. 4 Location of max rotation

Fig. 5 Rotation values along exterior girder G1 when rotation reaches maximum (Case 40)

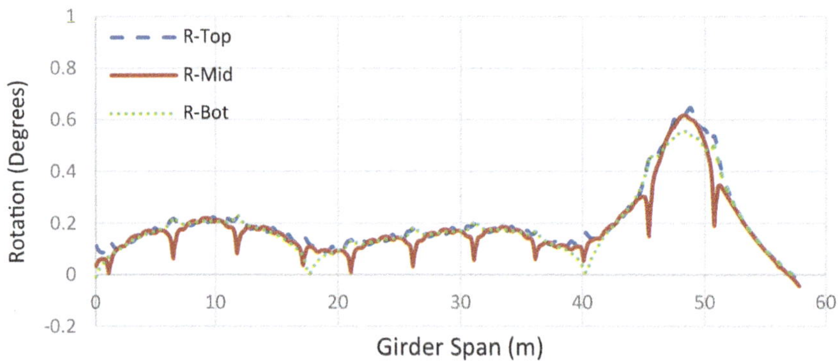

Fig. 6 Rotation values along exterior girder G6 when rotation reaches maximum (Case 43)

4.2 Skew Angle Effect

The skew angle of the bridge changed from the original 14° to 0°, then doubled to 28° and then tripled to 42°. It was noticed that the rotation in both exterior girders varied with the skew angle changes. However, the location of the maximum rotation in either exterior girder did not change. In fact, the results showed that with increasing the skew angle, the rotation of exterior girder (G1) increases, while the rotation of exterior girder (G6) decreases. The maximum rotation values in girders G1 and G6 that align with the different skew angles are illustrated in Fig. 7. The rotation of G1, when the bridge has 42° skew angle, is 31% higher than rotation of girder G6. This difference in rotation between the two girders is alarming and raises a lot of safety concerns.

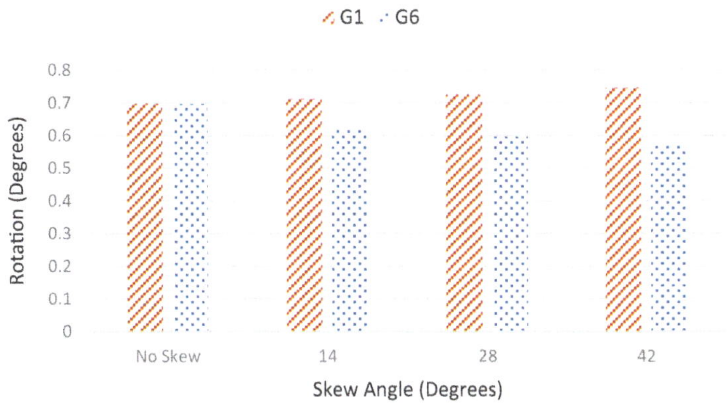

Fig. 7 Maximum rotation of exterior girders vs. bridge skew angle

4.3 B/D Ratio Effect

It was reported by Ashiquzzaman et al. that the ratio between the lateral bracing spacing (B) and the girders depth (D) is a good indicator to limit the rotation of the exterior girder [5]. Therefore, it is important to examine B/D ratio in skewed bridges. The maximum B/D ratio for Ford County Bridge is 7.8 which is larger than the recommended value of 4.0. Therefore, to reduce this ratio temporary bracing system is needed to reduce the distance (B). Diagonal pipe with a transverse tie (DP), horizontal pipe with a diagonal tie (HP) and intermediate cross-frame (CF) are three options of temporary bracing systems were considered in this study [1]. These systems were applied at the middle-distance between the intermediate diaphragms of exterior bays to reduce maximum B/D ratio to 3.9 as shown in Fig. 8.

Illinois department of transportation has a limit of 3/16th of an inch for the deflection of the overhang [6]. This deflection limit corresponds with a rotation limit for this bridge of 0.307°. Figure 9, shows that applying any of the

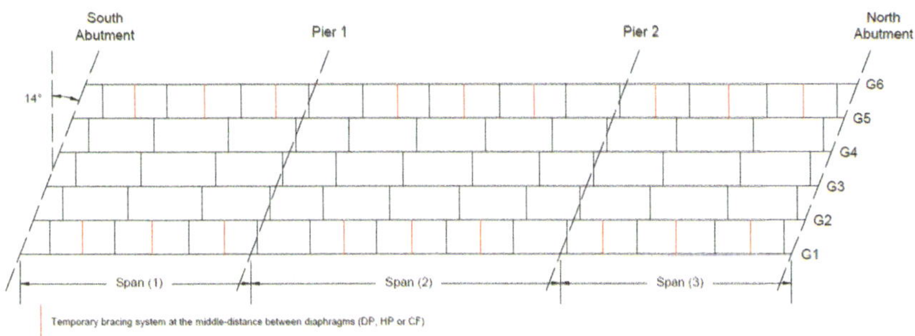

Fig. 8 Temporary bracing systems location

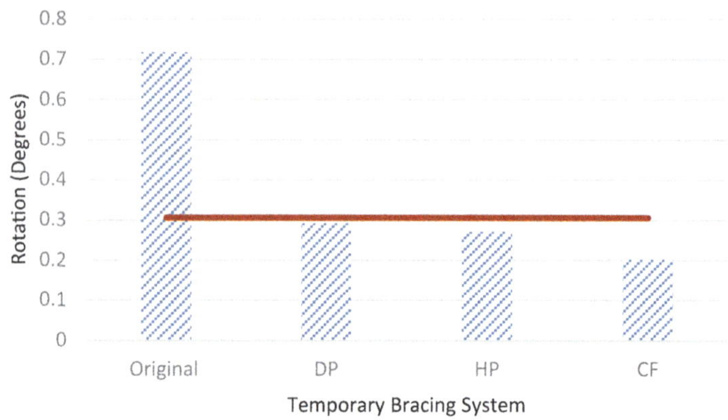

Fig. 9 Effectiveness of temporary bracing systems

recommended temporary bracing systems to limit B/D ratio below 4, reduces the maximum rotation of the exterior girders significantly, and keeps it below the limit.

5 Conclusions

The skew angle effect on the rotation of the exterior girders during construction was evaluated in this study. SAP2000 was used to develop Finite Element models for bridges with different skew angles and different temporary bracing systems. From the analysis and the comparison of the results the following can be concluded in skewed bridges under construction loads:

- Skewness of bridges leads to differential rotation between the two exterior girders of the bridge
- The differential rotation due to skewness increases when the bridge skew angle increases
- B/D ratio is a good indicator to limit the rotation of exterior girders in skewed bridges
- Adding Temporary bracing systems in the external bays of the bridge to reduce B/D ratio to is an effective method to reduce the rotation.

References

1. Ashiquzzaman M, Hui L, Schmeltz J, Merino C, Bozkurt B, Ibrahim A, et al (2016) Effectiveness of exterior beam rotation prevention systems for bridge deck construction. FHWA-ICT-16-015, Springfield, IL
2. Clifton SP, Bayrak O (2008) Bridge deck overhang construction. IAC 88-5DD1A003-2, Austin
3. Fasl JD (2008) The influence of overhang construction on girder design. University of Texas at Austin
4. Ashiquzzaman M, Calvo CM, Hui L, Ibrahim A, Lindquist W, Hindi R (2017) Effectiveness of different bracing systems to prevent exterior girder rotation during bridge deck construction. Eng Struct 142:272–289
5. Ashiquzzaman M, Hui L, Ibrahim A, Lindquist W, Thomson M, Hindi R (2016) Effect of inconsistent diaphragms on exterior girder rotation during overhang deck construction. Structures 8:25–34
6. Illinois Department of Transportation (2012) Bridge Manual. Illinois Departmentof Transportation (IDOT), Springfield, Illinois

Reconstruction of Partially Collapsed Post-tensioned Beğendik Bridge During Balanced Cantilever Construction

Alp Caner, Nurdan Apaydın, Melike Cınar, Erol Peker, and Mehmet Kılıc

Abstract Beğendik Bridge with a main span of 210 m has partially collapsed during construction in December 2017. The main reason for the partial collapse was a result of unfortunate events following each other. The collapse was simply triggered by removing a simple piece of timber support element from the scaffolding to have some space to place the unexpectedly oversize pot bearings on the abutment. No one has been injured or died during the collapse mainly due to ductile design of superstructure that allowed seven workers to escape from inside of the box girder within forty-five minutes. The focus of this paper on the description of the trigger mechanism of the partial collapse of the bridge, demolishment of damaged parts and reconstruction of the bridge. Following the collapse, a series of engineering scenarios have been evaluated to restart the construction. Reconstruction of the bridge has been started just in four months of time and bridge is still under construction.

1 Introduction

Begendik Bridge located at the South-Eastern part of Anatolia, has been partially collapsed due to a series of unfortunate events happening at the same time. Prior to getting into detailed investigation of collapse, it will be better to describe the

A. Caner (✉)
Middle East Technical University, Ankara, Turkey
e-mail: acaner@metu.edu.tr

N. Apaydın
KGM, Turkish Highways, Ankara, Turkey

M. Cınar
BridgeWiz, Ankara, Turkey

E. Peker
Burakcan Ins, Ankara, Turkey

M. Kılıc
AGM Muh, Ankara, Turkey

P. Gülkan et al. (eds.), *Developments in International Bridge Engineering*, Springer Tracts on Transportation and Traffic 17, https://doi.org/10.1007/978-3-030-59169-4_17

characteristics of the bridge to give a better picture for the readers. A general view of the bridge is given in Fig. 1.

The 450 m long bridge has three spans. The spans have variable depth with a maximum of 12 m at the face of pier and with a minimum of 4.5 m' depth at mid-span and at the two abutment sides. With a main span of 210 m, Begendik Bridge will be the longest one in its class in Turkey.

The box girder section has a width of 14 m allowing two lanes of traffic. The top flange of the box has two sidewalks for pedestrians. Top flange of the bridge accommodates traffic barriers for vehicles and railings for pedestrians as safety features. The concrete quality of the superstructure is set to 45 MPa.

The design of the bridge is based on modified AASHTO LRFD specifications [1]. The live load truck is taken as H30S24 truck that has been used for design of Turkish bridges (Fig. 2). The load factors and resistance factors in design are directly taken from the AASHTO-LRFD. An influence line analysis has been performed to determine the maximum effects under live load on the superstructure. The dynamic allowance on live load is based on AASHTO-LRFD specifications.

The earthquake design loads during construction at the cantilever position of the bridge superstructure is determined to be 0.12 g and design earthquake effect with 1000 year return period is taken as 0.57 g. It has been known that the nearest fault line to the bridge is about 10 km away from the construction site. Pier has been designed to remain essentially elastic and superstructure has been designed to remain elastic during the design earthquake.

Fig. 1 Rendering of bridge

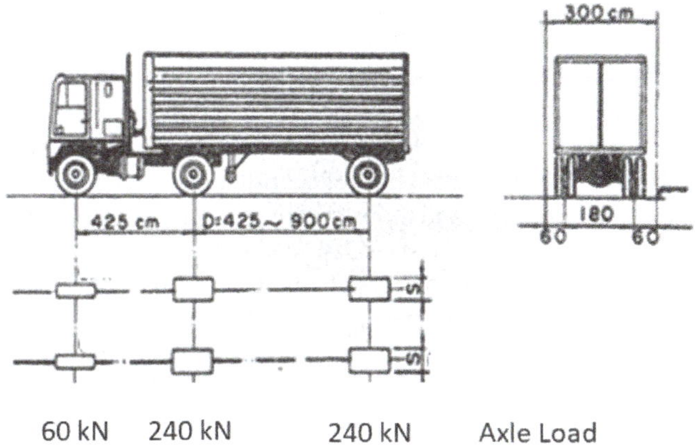

Fig. 2 H30S24 truck load model

2 Partial Collapse and Demolishment

The P1 pier has been constructed and balanced cantilever construction of the superstructure has been finished in December 2017. The last segments of the superstructure have been constructed over the temporary scaffolding to avoid unbalance force effects on pier P1 as shown in Fig. 3. The segment over the abutment has not been constructed and the pot bearings has not been placed. The pot bearings brought to the site were larger than the ones proposed in original design with some couple of months of delay and some of the scaffolding elements need to be deconstructed to

Fig. 3 Pier 1, superstructure and scaffolding

open space for pot bearing placement. At that moment, the scaffolding did not show any distress for couple of months.

Once some of the scaffolding elements were removed for bearing placement, the scaffolding has shown some signs of distress. The unbalanced moments let the superstructure to sag towards abutment and last segment before the abutment touched to the abutment wall with loss of support of scaffolding. At that moment only top post-tensioned strands are pulled and the bottom ones are not pulled due to not having the last abutment segment. Once the last constructed segment touched the abutment, positive moment developed along the superstructure and the superstructure is not fully ready for that state since the bottom post-tension strands are not even placed. The structure withstands overstress for about forty five minutes showing all signs of collapse. Seven workers escaped from inside of the superstructure. No body wounded or died during the collapse and luckily no one was passing the road underneath the bridge. Perhaps, the main reason for no one being died is the ductile design required by the seismic requirements. The superstructure has been designed with response modification factor of 1.0 as required by the AASHTO-LRFD specifications.

After the collapse, superstructure at abutment side span is measured as 60 m and the main-span side is 105 m as shown in Fig. 4. Even if most of the post-tensioned strands are damaged, the diaphragm section continued to carry the unbalanced moments. The unbalanced moment developed at the pier foundation is much less than the foundation design moment. Since the bridge has been designed for an extreme earthquake, the structure remained in elastic state for about two months with a drift less than elastic state limit at top of the pier. Everyday elevation readings have shown no sign of yielding and movement during this two months' time.

It has been observed that the remaining superstructure is unrepairable and pier P1 and its damaged superstructure can be demolished. The work safety requirements on superstructure cannot be maintained and forced us demolishment of pier at ground level. In demolishment of pier P1, dynamites have been placed about 20 m from the foundation level as shown in Fig. 5. It has been well known that the

Fig. 4 Unbalanced structure remained in elastic state

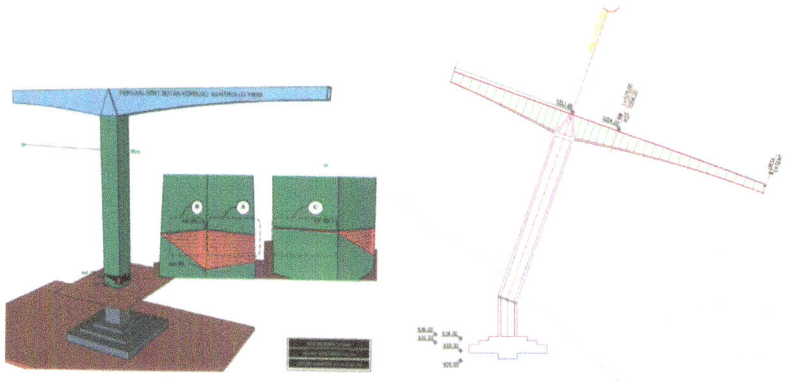

Fig. 5 Demolishment plan

dynamite system cannot cut the bars with explosion. The dynamite pattern has been redesigned to release of longitudinal reinforcement from the concrete. In redesign, additional dynamites have been placed on tension face where there was only one row of dynamites. Otherwise, the whole foundation will be pulled by reinforcement and foundation can get damaged.

During the fall down of structure to ground, It has been estimated that the structure can develop a 0.1 g peak ground acceleration during demolishment. The measured acceleration on ground was determined to be 0.1 g. The pier has fell into the valley as planned. The structural analysis indicate that the foundation has not been damaged since the developed forces during explosion is much less than the design loads set for extreme earthquake event.

The superstructure is estimated to crush under compression as the structure hits to the ground since the concrete reaches its crushing stress as shown in Fig. 6. As predicted, the dynamite pattern allowed unbonding of longitudinal reinforcement from the remaining pier concrete at elevation 20 m as shown in Fig. 7. The pier itself was just laying on the ground as predicted not far away from the pier P1 foundation.

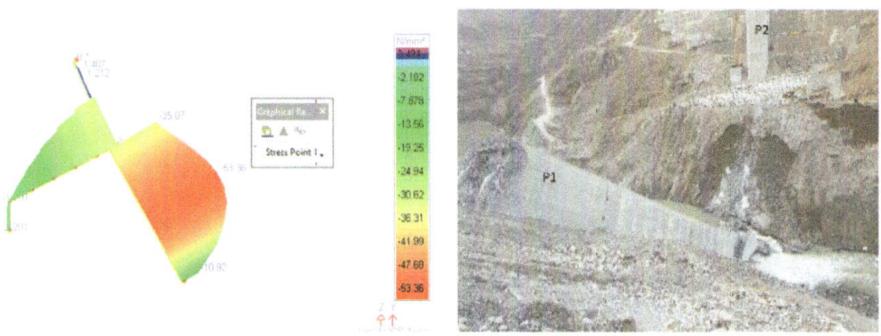

Fig. 6 Structural predictions and observations after demolishment

Fig. 7 Pier 1 remaining concrete after demolishment

3 Reconstruction

Top of pier P1 needs to be cleaned up for reconstruction of the pier. A six-meter region was cut from the top to start reconstruction of pier P1 as shown in Fig. 8.

From the remaining part shown in Fig. 8, samples for concrete cylinder tests and rebar tests has been taken. The results of the cylinder tests indicate that the concrete has a crushing strength of 55 MPa. The tensile strength of the rebars reached 575 MPa. The material tests indicate that the results are satisfactory in terms of design and reconstruction.

Even if the section is adequate to carry the design loads, a concrete jacketing was designed to eliminate any unforeseen weaknesses as shown in Fig. 9. The concrete jacketing is only covers ±5 m from cut section and helps to transfer mainly moment and shear forces on a larger section. The concrete jacketing is based on seismic retrofit manual for highway bridges [2]. The pattern of mechanical connectors used for longitudinal bars were selected based on AASHTO-LRFD [1].

On pier P2 side, the construction of the cantilever part has almost been completed as shown in Fig. 10. This pier is very similar to the Pier P1. The complete construction stages have been studied once more and determined that the construction stage stresses are less than the ones indicated in the limits as shown in Fig. 11. Different

Fig. 8 Cleaning up top of pier 1

Fig. 9 Design of concrete jacketing and addition of longitudinal bars

Fig. 10 Pier P2 under construction

construction scenarios have been studied to decide on closure of the cantilever part to the abutment segments. An optimum solution has been determined. The review of original design computations is found to meet design requirements.

On site, post-tensioned strand jacking forces and elongations are controlled by couple of ways. Elongation control helps to determine more realistic wobble coefficients at the site rather than the ones suggested by the specifications. Wobble coefficient simply varies from one duct to other based on the straightness of duct which can not be perfectly maintained during construction due to fresh concrete placement. An open source calculator for free use is also available as shown in Fig. 12. The calculator has been used at the construction site for elongation verifications and to assess

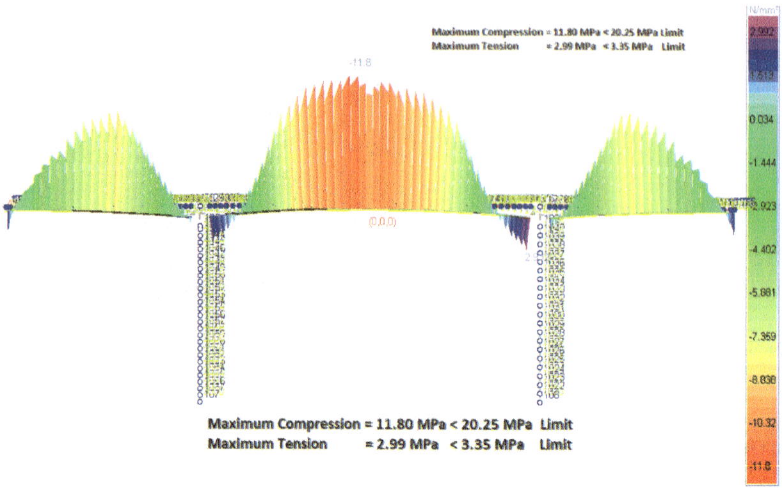

Fig. 11 Computer model [3] and maximum stresses

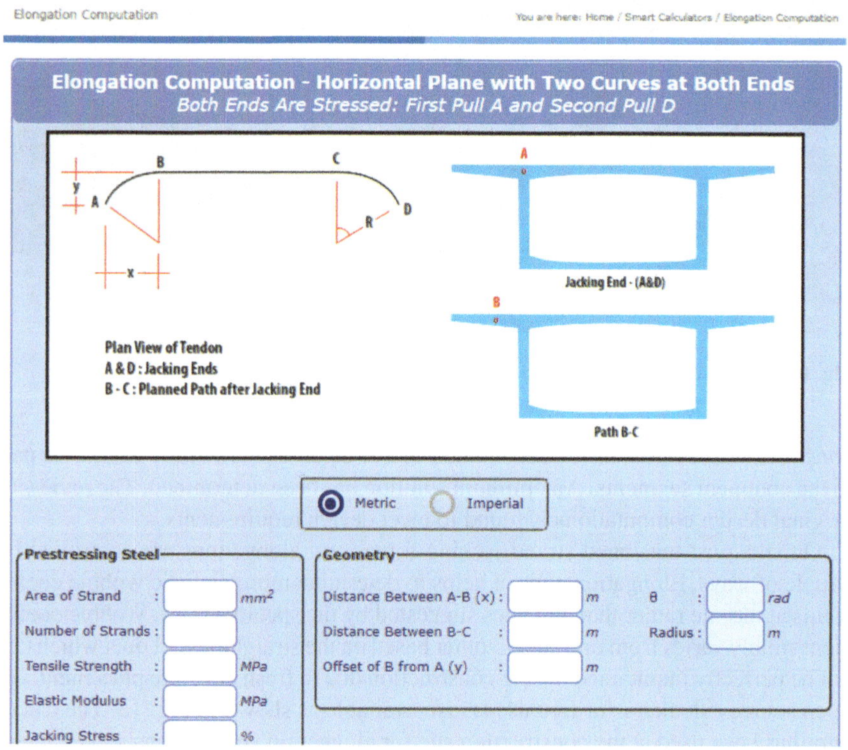

Fig. 12 Open source free calculator for elongation control [4]

the more realistic wobble coefficients. It shall be noted that the wobble coefficient of the specification can be sometimes ten times larger at the construction site.

In addition, the thermocouple readings were made along the depth of the structure to determine the effect of thermal gradients as shown in Fig. 13. Maximum gradient between the top surface and inner face is determined to be around 7 °C maximum. It has also been observed that the temperature gradient also changes not only in vertical direction but also in transverse direction based on position of the sun facing the girder as shown in Fig. 13. The measured thermal gradient is less than design values. Average measured relative humilities were about 20%, an indication of more creep and shrinkage strains on the structure compared to the standard case of 70% relative humidity.

The reinforcement detailing for blister has been shown in Fig. 14. The concreting and curing has been shown in Fig. 15. It shall not be forgotten that the wet blanket will protect concrete at low humidity climate from premature cracking. For any bridge

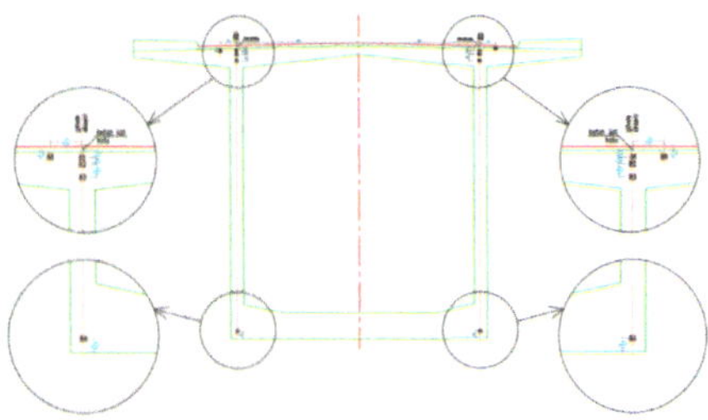

Fig. 13 Thermocouple locations and measured thermal gradient

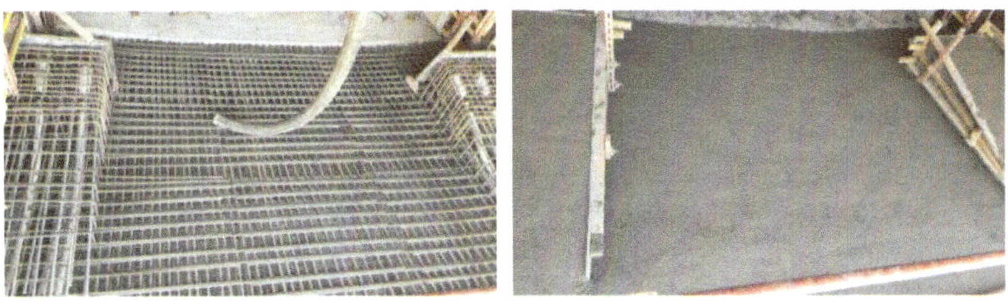

Fig. 14 Blister reinforcement and formwork

Fig. 15 Curing concrete after placement

construction, characteristics of time dependent effects such as creep and shrinkage can be totally different based on the humidity.

4 Conclusion

The following conclusions can be withdrawn from this paper.

1. Partial collapse of the bridge is a result of removing elements from scaffolding to place the unexpectedly oversize pot bearings over abutments that lead to collapse of scaffolding.
2. The ductile response of the superstructure gave about fortyfive minutes of escape time for workers on the bridge. It shall also be noted that the superstructure was designed for an extreme earthquake event that required additional rebars in the superstructure that helped to improve the ductile response. For two months, the unbalanced structure show no sign of movement or yielding prior to demolishment.
3. Pier P1 has been demolished as planned using a proper dynamite pattern and no damage was given to the foundation. During the fell down of structure, the ground vibrations are much less than the extreme earthquake event.
4. Even the material tests of samples taken from damaged zones showed that the minimum required design values are satisfied.
5. A concrete jacketing was provided to have additional strength at the interface of existing and new concrete at pier P1.
6. The construction of the structure has been monitored through site visits and additional measurements such as thermal gradients and post-tension elongation. Site concrete creep tests will be performed soon.

Acknowledgements The authors would like to thank to Turkish Highways for their support. The authors would like to extend their thanks to Burakcan Ins and AGM Muh for their significant help.

References

1. AASHTO (2010) Bridge design specifications, 5th edn., Washington DC
2. FHWA (2006) Seismic retrofitting manual for highway bridges, part-1 bridges, McLean VA
3. LARSA (2018). https://www.larsa4d.com/products/features/4dbridge-bridgeplt.aspx
4. BridgeWiz (2018). https://bridgewiz.com/calculator/elongation/elongation.html

Construction of Namawukulu Footbridge in Uganda

Sercan Durukan and Xavier Echegaray

Abstract For many years, Bridges to Prosperity has been trying to improve people's lives in the isolated communities throughout the world by building footbridges. In this concept, recently Ramboll UK and IABSE Foundation has teamed up to realize a suspended footbridge project located in the eastern Uganda. A multi-national team of structural engineers travelled to site to construct the bridge together with the local community. In this paper, the suspended bridge design that enables the use of basic materials and the construction methods developed considering the available limited resources is presented.

1 Introduction

Bridges to Prosperity (B2P) is an award-winning Non-Governmental Organization (NGO) committed to tackle rural isolation by delivering safe, sustainable and lasting footbridges worldwide. These structures do not only provide safe crossings over mighty rivers during the rainy season, but also connect rural communities to essential services and infrastructure, laying the foundations for progress and prosperity.

So far, this United States—based NGO has built over 200 footbridges across the globe in liaison with local governments and stakeholders and relies on the support of numerous partners across the AEC industry. Its operation focuses in Central and South America, Africa and South East Asia, where they have recently reached the impressive mark of one million bridge users.

One of Bridges to Prosperity's core belief is that "rural isolation is a root cause of poverty", and that "connection is the foundation for opportunity". A study by the University of Notre-Damme found that B2P's reliable river crossings increased household income, farmer crop yields and employment rates by approximately 30%. The World Bank has estimated that over one billion people lack permanent access to

S. Durukan (✉)
Bridge Engineer, Bridgewiz, Istanbul, Turkey

X. Echegaray
Bridge Engineer, Ramboll UK, London, UK

P. Gülkan et al. (eds.), *Developments in International Bridge Engineering*, Springer Tracts on Transportation and Traffic 17, https://doi.org/10.1007/978-3-030-59169-4_18

basic infrastructure. With their footbridges, B2P aim at eradicating poverty caused by rural isolation, providing communities in need with an all-year-round vital connection to basic services like healthcare and education and opening the door to further economic opportunities.

Bridges to Prosperity has established a long-term relationship with the African continent. The Ugandan government has taken great interest in B2P’s work and has recently opened the door to hosting and supporting a ‘Country Program’ within its borders. The project object of this paper is part of this exciting and ground-breaking initiative, as it is the first industry-sponsored bridge in Uganda.

Through the Industry Partnership program, Ramboll UK and IABSE Foundation have recently cooperated to secure the essential funding and provide volunteers to travel to this specific bridge site in Uganda.

2 Project Site

The Namawukulu footbridge is located in Bukalasi, within the Bududa municipality of Eastern Uganda. The site is within the Ugandan part of the pristine Mount Elgon National Park, and is less than 10 km away from the Kenyan border with the GPS Coordinates as: 1^{0}00’16.2”N 34^{0}24’06.0”E.

Bukalasi and its surroundings host the Mamawukulu, Bumbian, Kishambua, and Malandu communities, which are dedicated to farming of vegetables and coffee as main means of subsistence. It is estimated that the bridge will directly benefit 1,300 members of these communities, with almost 800 of them being children.

The Ngame river crosses an existing footpath that connects these communities to basic services and institutional centers, including three health clinics, two schools, training centers, the local market and the government office amongst others. This means that when the river is too high to cross during the rainy season (approximately 30 days a year), local residents remain isolated. This not only has a direct negative impact on farmer trade, household income, the education of the younger ones, but

Left bank of the Ngame river   Existing crossing method

Fig. 1 Bridge location view during dry-season

also poses a significant threat to the lives of those that may require immediate access to healthcare.

The Namawukulu footbridge will provide an all-year-round safe alternative to an existing rock 'crossing', which even in the dry season can be life-threatening considering that the majority of local community members cannot swim (Fig. 1).

3 Design

Equally important to identifying and understanding the need for connection is delivering the right solution. The design has to be developed, optimized and perfectioned to make it fit for purpose. The suspension bridge design provided can withstand extreme climatic phenomena (including seismic events) and heavy pedestrian, animal and light vehicle traffic. To provide a long-lasting solution, the design of structural details and the choice of materials are focused on making these bridges easy to maintain and durable as well as enabling use of local labor and materials whenever possible.

This extensive design intent has been captured and codified in Bridges to Prosperity's 'Bridge Builder Manual' [1]. To make the design process more efficient, replicable and accessible, B2P has developed a set of standardized designs that cater for different bridge geometries and site conditions. This approach eliminates the need of performing detailed analysis for each bridge and provides greater certainty and background to designers without proficient engineering knowledge. The manual however has its limitations, and as with any standardized design approach, engineers need to use their judgement to determine the applicability and adequacy of the codified assumptions to the conditions on site.

The Namawukulu footbridge spans 70 m over the Ngame river. As a suspended footbridge, its cables work in pure tension and support the deck between anchorage points. The deck is naturally curved in elevation under distributed vertical loads, following the shape of the catenary. At the abutments, the cables are supported on vertical towers and anchored at the far bottom-end of the foundation [2] (Fig. 2).

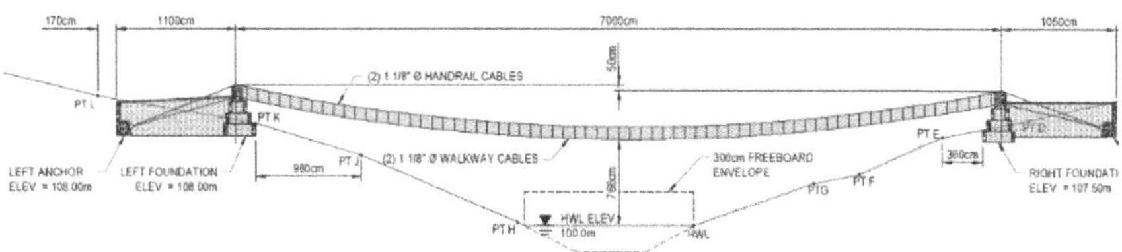

Fig. 2 General arrangement of the bridge

3.1 Design Basis

Although most design codes follow a widespread semi-probabilistic method based on the use of partial factors (with distinct factors for material strength and actions) and limit-state performance, Bridges to Prosperity's 'Bridge Builders Manual' uses the Allowable Stress Design (ASD) methodology to determine structural reliability, which has proved to be safe and practical.

The ASD method limits the elastic stress in a structural element under service loading. The structural verification of an element relies on the ratio between the nominal limit stress (Rn) and the stress in the element (Qi) being greater than a predetermined factor of safety (FS). The factor of safety considers the criticality of the element and the uncertainty involved in determining its resistance and imposed load effects.

$$\sum Q_i \leq \frac{R_n}{FS}$$

As opposed to the partial factor method and the use of limit-state performance, ASD considers that all load actions have the same variability thus making combinations and analysis slightly simpler. This is in line with the 'Bridge Builder Manual' objective of making the design process more efficient.

The material properties used in design is presented in Fig. 3 below.

Note that geotechnical properties have been determined based on data collected on site.

The load effects considered in the design is also summarized below:

Dead Load = 0.87 kN/m.
Uniformly Distributed Live Load = 3.33 kN/m.
Wind Load = 0.50 kN/m.
Point Live Load = 2.25 kN.

Note that the point load represents potential livestock loads and lightweight traffic loading.

UNIT WEIGHTS:			MATERIAL PROPERTIES:			
STEEL	=	7850 kg/m^3	CONCRETE	f'c	=	10 MPa
CONCRETE	=	2300 kg/m^3	REINFORCING	fy	=	275 MPa
TIMBER	=	900 kg/m^3	TIMBER	fb	=	3.96 MPa
BROKEN ROCK	=	2000 kg/m^3	TIMBER	fv	=	1.44 MPa
MASONRY	=	2100 kg/m^3	SOIL	qa	=	144 kPa
SOIL	=	1800 kg/m^3	FRICTION ANGLE	Ø	=	33 degrees

Fig. 3 Material properties considered in design

3.2 Geometric Design

Following the design guidelines, the foundations and the deck must respect a series of geometric constraints. Ultimately, these restrictions determine the relative location of the bridge abutments and subsequently, the span length of the bridge.

To begin with, the abutments must be located with a minimum setback of 3 m from the edge of the river bank and behind the line of the angle of soil internal friction (35°) of the bank. Furthermore, if the abutments lie on soil, its slope should be reduced to less than 10°. These measures are aimed at eliminating erosion and slope stability problems that could potentially affect the abutments during the lifespan of the bridge.

One of the most important constraints that may influence the position of the foundations is the deck freeboard. To avoid a catastrophic collision of debris carried by the river against the bridge during flooding events, the deck should provide a minimum clearance of 3 m above the river's high water level (HWL). In the case of the Namawukulu bridge this constraint is not critical, as the deck soffit is 7.6 m above HWL.

The distance in plan between the abutments will determine the span length, which must be under 120 m so that aerodynamic effects and wind stabilization measures can be ignored. Once determined, the sag of the cables under permanent loads can be calculated as 5% of the span length (please refer to fig??). This percentage of design sag is a compromise between serviceability and structural capacity. A small sag to span ratio results in shallow slopes at both ends of the deck, enhancing user comfort. However, a straighter deck with shallow slopes at both ends leads to high forces in the cable, increasing structural demand on the foundations.

For this same reason, the difference in height of the cable saddle supports and the foundations is limited to 4% of the span length, as having one support higher than the other leads to higher slopes at the high end of the deck and increased eccentric forces imposed on the cable support tower at its lower end.

The deck is 1.0 m wide to allow for pedestrian, livestock and light vehicular traffic. Although the design could be engineered for a wider deck, this is not implemented to avoid the uncontrolled access of heavy vehicles.

3.3 Cable Analysis

Cable geometry (I have included the f-value stuff in the setting the cable part).

Cable design checks for resistance based on tension (F.S. = 3).

** If you wish to add sth else I think we can trim the design intro or the part with ASD vs LRFD.

4 Construction

As expected, the bridge was located in a highly remote area and available material, equipment and labor resources were very limited. In order to construct the bridge without compromising safety and quality, site specific construction methods required to be developed. In this chapter, an insight will be given about the important methods used to finish the construction in 12 days.

By using common and readily-available materials and components, this bridge design is constructible in any country using local resources and labor. The only component that cannot be easily sourced locally are bridge cables. These are usually sourced from cargo ports, which donate used cables from gantry cranes. Overall, the construction model is replicable, efficient, and sustainable, as it minimizes transport and upcycles used material. A very good example to this is presented in Fig. 4 which shows an automobile wheel used as a cable saddle on towers.

The substructure was constructed using conventional methods by the local community before the volunteer team has arrived therefore this chapter will cover superstructure construction activities only.

4.1 *Setting the Cable Sag*

Naturally, the cables are the most crucial element in the structural system of this suspended footbridge. When the cable is positioned in its place and hanging between the towers, it takes the funicular form of catenary.

Sag is the vertical drop of the cable at the mid-point. When the two anchorages are at the same level, the mid-point will be the lowest point on the cable but if the towers have a difference in level (ΔH), then mid-point will not be the lowest point. For convenience, cable position needs to be checked using the lowest point because it is possible to identify it from any angle with the auto-level. For this purpose, the f-value is used which is the height difference of between the lower tower saddle and the lowest point of the cable (Fig. 5).

The f-value is calculated from the below formula:

$$f = \frac{(4 \times h_{hoist} - \Delta H)^2}{16 \times h_{hoist}} \quad (1)$$

Fig. 4 Geometric constraints

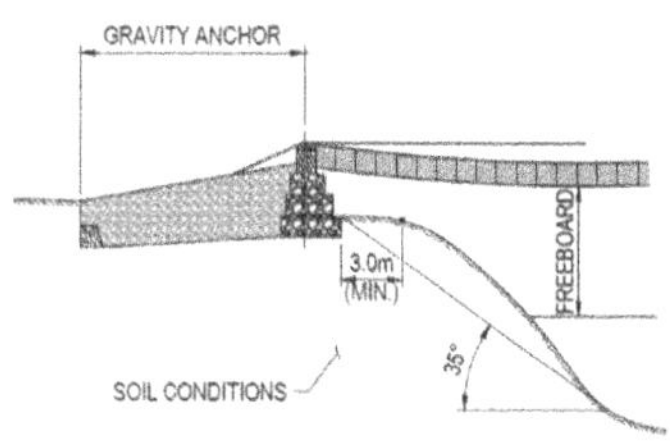

Fig. 5 The wheel cable saddle

The above formula can be used to convert any sag value to an f-value. The cable will be adjusted to an initial hoisting sag which is smaller than the final state so that when all the dead load is placed the bridge will achieve the design sag of 5%. The hoisting sag is calculated as 4.54% which means:

$$h_{hoist} = 0.0454xL \tag{2}$$

First of all, the as-built dimensions of the towers are taken and the as-built ΔH is measured which was 1 cm off from the design drawings and the f value is calculated as 2.94 m and hoisting sag as 3.18 m. The f-value is marked on the tower and the auto-level is set to this mark so that the cable level could be checked accordingly.

All of the four cables (2 handrail & 2 walkway) need to be winched up to achieve the above calculated sag. The cable has a dead anchorage on higher side and an active anchorage on the lower side. The entire winching operation is performed in the active anchorage (see Fig. 6).

Basically, the live end of the cable is hooked and attached to the winch cable as shown below in Fig. 7. Winching is done until the cable goes slightly above the f-value and then all the live end is clamped to the dead end. The fine tuning of the cable level is done through releasing some clamps and vibrating the cable so that the cable can slip slowly under control and positioned in desired level. The walkway cables and handrail cables have to be leveled to each other as well.

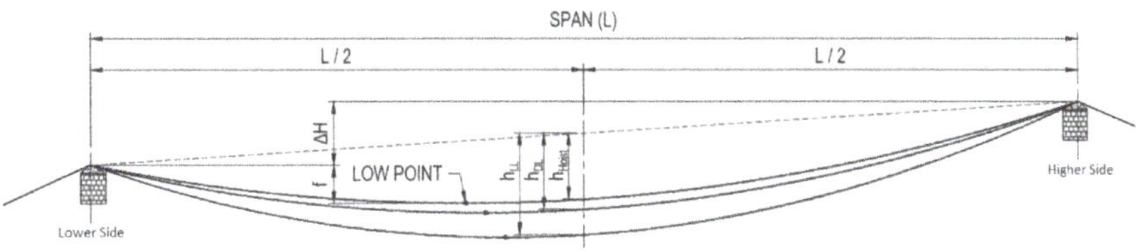

Fig. 6 Cable hoisting sag and f-value

Fig. 7 View from active anchorage and the auto-level set up

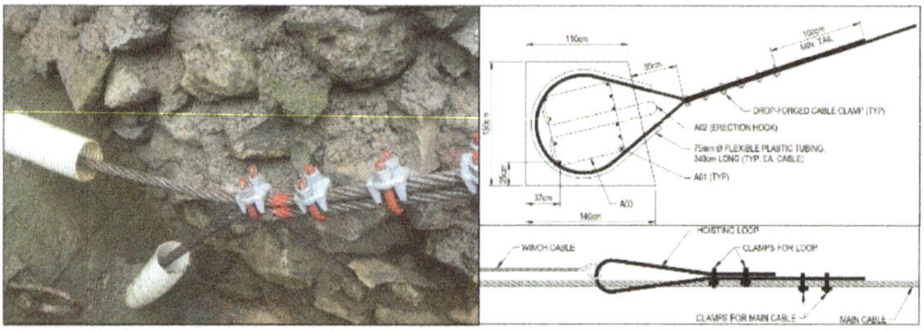

Fig. 8 Cable termination detail at the anchorage with the spray paint marks and the hoisting hoop

It is crucial that the entire cable adjustment operation is performed under same weather conditions because the cable sag is highly affected by the temperature due to thermal expansion of the cable. In this view, the cable adjustment job is performed early in the morning.

After getting the cable at the desired level and sag, 24 h of observation period is allowed to makes sure that cables keep the position. The cables are also marked with spray paint at the anchorage to check if there is slipping (Fig. 7). If needed, re-adjustment is made after the observation period and the cable is fixed to its final condition by using 6 drop forged clamps with 15 cm spacing. The clamps are torqued until 25% of reduction is observed in the cable diameter. The cables are then coated with tar and the cable ducts in the anchor block is grouted for corrosion protection (Fig. 8).

4.2 Cross Beam Pre-fabrication and Erection

The bridge timber decking will be resting on 1 m spaced cross beams on the walkway cables. In order to speed up the construction, the cross-beam assembly is pre-fabricated at a workshop in site. The cross-beam assembly is composed of the steel channel profile cross beams, the timber nailer boards which will bond the timber decking to the steel cross beams and the suspenders which are 10 mm deformed bars.

Fig. 9 The crossbeam assemblies

At the shop, the crossbeams are painted, the nailer boards are drilled and attached to the cross-beams and the suspenders are bent to form a hook at the top side (Fig. 9).

The cross-beams are located in every 1 m of the bridge but it is not possible to access to the most part of the cable and there is no equipment to lift up the heavy assembly. Therefore, they need to be launched from the abutment sides. The volunteer team developed a unique system to ease this task with the available materials. All of the crossbeams stacked on the higher abutment side and they are tied together with a rope. The knots on the rope are spaced 1 m also so that when the rope is pulled from the other side, all off the cross beams will be positioned correctly. This idea of pulling the heavy cross beams like a curtain has been applied very successfully and now it is being used in other Bridges to Prosperity projects.

4.3 *Finishing Works*

After all the cross beams are positioned, the longitudinal timber decking boards are attached on top of the nailer boards. This operation can be tricky because the used timber boards are not perfectly cut so care should be taken to keep the straight alignment of the boards.

During the drilling and bolting of the decking plates, a separate team has worked on pre-fabricating the fencing units in order to open the bridge on time. The wire mesh used for fencing needs to be stretched when it is assembled to the deck. For this purpose, the wire mesh is stretched on the ground and nailed to a timber curbs on the bottom. This way the mesh can remain stretched and could be attached to the decking board directly. After the fences are fixed to the deck, the mesh is bent over the handrails and tied with wires. (Fig. 10). The anchor pits are filled with rocks and the concrete topping is casted above them.

After all the construction activities finished, maintenance training is given to the community members, some of who already involved in the construction on purposely by Bridges to Prosperity to transfer the skills and knowledge to the other members of the community.

Fig. 10 Views of the finished bridge

When the inauguration has arrived, the community celebrated to have this bridge for whole day. Too see hundreds of people crossing the bridge by walking and riding boda-bodas over it, experiencing and experimenting the bridge with grear amusement worth all the effort.

Acknowledgements The authors, would like to thank Bridges to Prosperity, Ramboll UK and IABSE Foundation for making the project possible and the volunteers Eduardo Leach, Tim Prosser, Wojciech Szewczak, Ed Fraser, Spas Ivanov, Alberto Turco, Sabrina Taffarel for volunteering and putting intense effort to finish the construction on time and companies Bridgewiz, Bolina and Schimetta for providing volunteers for the IABSE Team.

References

1. Bridges to Prosperity Bridge Builder Manual, 5th Edition, Denver CO, USA
2. Bridges to Prosperity - Namawukulu Suspended Bridge Design Drawings, Denver CO, USA

Life-Cycle Environmental Impact Assessment of Steel Bridge Deck Pavement

Xiang-fei Zhang, Zhen-dong Qian, and Hui Gao

Abstract To evaluate the environmental impact of steel bridge deck pavement (SBDP), based on life cycle assessment (LCA), this study quantified energy consumption and gas emission of epoxy asphalt (EA) mixture, stone mastic asphalt (SMA) mixture and guss-asphalt (GA) mixture. Firstly, the life cycle inventory of SBDP materials was established, involving the stages from raw material acquisition to end of life. Subsequently, environmental impact assessment indicators were proposed to evaluate energy consumption, climatic change and human health. Secondly, the uncertainty assessment method was investigated, and the uncertainty of inventory data and environmental factors was analyzed subsequently. Thirdly, "EA + EA" structure and "GA + SMA" structure were analyzed based on the established LCA model. Results indicate that the environment impact of "GA + SMA" is greater than "EA + EA", and the stages with high environmental impacts of the two typical structures are the raw material acquisition, plant production, and operation and maintenance, successively. In addition, the production of epoxy asphalt binder and the plant production of GA and SMA should be optimized to reduce energy consumption and gas emission.

1 Introduction

Steel bridge deck pavement (SBDP) is an important part of pavement engineering, and should pay attention to and promote energy-saving and environment-friendly technologies. During the construction and maintenance process of the pavement,

X. Zhang (✉) · Z. Qian
Intelligent Transport System Research Center, Southeast University, Southeast University Road, Nanjing 211189, China
e-mail: zhang_xf@seu.edu.cn

Z. Qian
e-mail: qianzd@seu.edu.cn

H. Gao
Organization Department of CPC Suyu District, 1 Shaoshan Road, Suqian 223800, China

P. Gülkan et al. (eds.), *Developments in International Bridge Engineering*, Springer Tracts on Transportation and Traffic 17, https://doi.org/10.1007/978-3-030-59169-4_19

application of asphalt mixture consumes great quantities of energy and discharges lots of gas. It is necessary to study the environmental impact analysis model of SBDP technology to reduce the energy consumption and environmental impact of the whole life of the engineering.

Life cycle assessment (LCA) provides a comprehensive and holistic platform for environmental assessment of sustainable practices [1]. It has been extensively used in different studies to assess environmental impact of pavements. Three different approaches are adopted to conduct an LAC, including process LAC, economic input–output LAC, and hybrid LAC [2]. To date, the application of LCA in SBDP engineering is obstructed by data credibility insufficiency and differences between SBDP and highway. Firstly, due to the imperfection of the basic data of asphalt energy consumption and the diversity of asphalt production and construction technology, LCA results are not corresponding with real environmental impact levels [3]. However, the credibility assessments of data quality and analysis result have not been involved in the associated studies of LCA in pavement engineering. Actually, achievements have been made in data quality assessment and analysis result evaluation in the studies of LCA methodology. The uncertainty index and sensitivity index were proposed utilizing data quality indicator method, Monte Carlo Simulation, and Taylor series expansion method [4]. Secondly, the calculation model of highway LCA cannot be directly applied to SBDP. The material production process and construction process of SBDP differ from highway pavement greatly [5]. Moreover, maintenance technology of asphalt pavement is not totally applicable to SBDP due to strict use conditions, complex stress states, and unique pavement structures [6]. Rare efforts have been made to investigate environmental impact of SBDP based on LCA.

The purpose of this study is to assess the environmental impacts of SBDP based on LCA. Specially, the environmental performance of the following three mixtures, which are commonly adopted as pavement materials in SBDP, were evaluated and compared: stone mastic asphalt (SMA) mixture, gussasphalt (GA) mixture, and epoxy asphalt (EA) mixture. A life cycle inventory (LCI) was established to quantify the energy consumption and gas emission during each life stage. Subsequently, environmental impact factors were computed based on LCI, and data quality and analysis credibility were evaluated utilizing the uncertainty index and the sensitivity index. Finally, the environmental impact analysis model was applied to investigate the environmental impacts of two general SDBP structures, which are "EA + EA" and "GA + SMA".

2 LCA Model for Steel Bridge Deck Pavement

The process LCA method was used to assess environmental impacts of SBDP. The results were compared between pavement structures, which were "EA + EA" and "GA + SMA" (Fig. 1). The system boundary includes raw material acquisition, plant production, on-site construction, operation and maintenance, and end-of-life

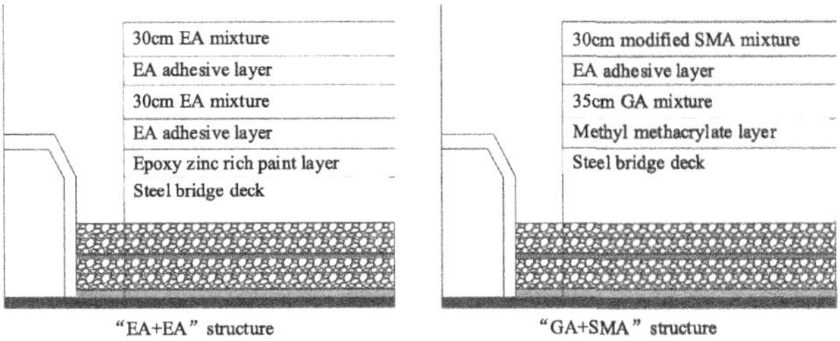

Fig. 1 SBDP structure composition

Table 1 Constituent materials

Material	GA binder	SMA binder	EA binder	Aggregate	Mineral powder
EA mixture/t	–	–	26.475	366.488	40.725
SMA mixture/t	–	26.250	–	392.963	43.65
GA mixture/t	37.615	–	–	336.488	106.275

stages. Maintenance treatments were developed based on existing researches [7]. The function unit was defined as 500 m semi-range deck with 11.25 m width. Considering the requirements of SBDP in Chinese technical specifications JTG D64-2015[8], 15 years was chosen as design service life in this study. The constituent materials of related mixtures are shown in Table 1.

The LCA approach first calculates the LCI of energy consumption and gas emissions. The data sources of LCI for material and production stages in initial pavement construction and maintenance were obtained from sources that met criteria for quality and relevance.

Raw Material Acquisition (RMA)

Raw materials of asphalt mixture include natural aggregate, mineral powder, virgin asphalt binder, polymer as asphalt binder additive and interlayer adhesive material. The main environmental impact data of aggregates and mineral powder and were acquired from Ecoinvent database [12]. Data related to petroleum asphalt were obtained from the investigations of European Bitumen Association (EBA) [13]. Trinidad Lake Asphalt (TLA), used in GA binder, can be applied to mixture only after filtering, and the environmental impact was neglected in this study. EA binder is produced by mixing two ingredients. Ingredient A consists of epoxy resin, and the relevant data was obtained from Ecoinvent database. Ingredient B consists of curing agent, asphalt and additive. The relevant data were obtained by investigating epoxy asphalt manufactures in Zhenjiang, China [14]. Fiber stabilizer in SMA and interlayer adhesive material are less than 1% of product weight, herewith the upstream data could be neglected [15].

Plant Production (PP)

During this stage, the energy consumption mainly originates from fuel burning in the heating process and electric power consumption in the operation of the equipment. According to the study of Prowell et al. [16], the heating energy required for aggregate drying is calculated using a thermodynamics based approach. The energy consumption of SMA and GA is acquired according to Prowell. Electric heating power of EA binder facilities is 29.5 kW for ingredient A and 11.5 kW for ingredient B according to the field research [14]. The gas emission mainly originates from fuel combustion and aggregate stacking, and were acquired from Ecoinvent database. The emission caused by electric power production (EA binder) is relatively complex, which is not involved in this study.

On-site Construction (OC)

The on-site construction stage is the most intensive stage in which transportation vehicles and construction machinery are used. LCI data of transportation was acquired from Ecoinvent database for GA and SMA mixture and Chinese Life Cycle Database (CLCD) [17] for EA mixture. Energy consumption of construction machinery of SMA and EA are familiar with ordinary asphalt pavement. Energy consumption of GA construction machinery was obtained from EBA [13]. The gas emission data was acquired from Chinese industry standard [18].

Operation and Maintenance (OC)

The operation and maintenance stage mainly includes vehicle operating and pavement rehabilitation. Due to vehicular fuel consumption rate is affected by multi-aspects, the vehicle operating is not considered in this study. Micro-surfacing is the main method of SBDP maintenance [7], and the relative data were obtained from EBA [13].

End of Life (EL)

Since most pavement materials have different levels of deterioration at the end of life, it would be appropriate to use the cut-off method for end-of-life allocation [19]. In order to protect the steel bridge deck, the pavement is initially removed by construction equipment leaving behind a pavement layer with the thickness of 25 ± 5 mm. Subsequently, the pavement layer is removed artificially, and the environmental impact was neglected in this study.

To investigate impacts on energy consumption, climate change and human health, the assessment indicator (*AI*) is limited to primary energy demand (*PED*), global warming potential (*GWP*), acidification potential (*AP*), and respiratory inorganic (*RI*). Due to some environmental factors have similar effects on the environment, the various indicators was classified and characterized. Influence factors (I_i) and characterization factors (C_i) are shown in Table 2. *AI* can be calculated through Eq. 1. *AI* of SBDP structures is shown in Table 3.

$$AI = \sum_i (Ii \times Ci) \tag{1}$$

Table 2 Influence factors and characterization factors

Impact category	AI	I_i	C_i		Factor sources
Energy consumption	PED	Fossil fuel	1	MJ	
Climate change	GWP	CO_2	1	kg CO_2 eq/kg	IPCC2013 [9]
		CH_4	28		
	AP	SO_2	1	kg SO_2 eq/kg	CML2002 [10]
		NO_x	0.7		
Human health	RI	PM	0.157	kg PM2.5 eq/kg	IMPACT2002 + [11]
		CO	0.001		
		SO_2	0.078		
		NO_x	0.127		

Table 3 Environmental impact assessment indicators of SDBP structures

AI		RMA	PP	OC	OM	EL	Life cycle
PED (MJ)	GA + SMA	1.97E + 06	8.50E + 05	8.52E + 04	3.20E + 05	2.19E + 04	3.25E + 06
	EA + EA	1.74E + 06	3.32E + 05	4.87E + 04	3.20E + 05	1.70E + 04	2.46E + 06
GWP/ kg CO_2 eq	GA + SMA	6.45E + 04	7.68E + 04	7.10E + 03	8.04E + 03	1.82E + 03	1.58E + 05
	EA + EA	6.34E + 04	2.23E + 04	4.06E + 03	8.04E + 03	1.42E + 03	9.93E + 04
AP/ kg SO_2 eq	GA + SMA	1.76E + 02	2.10E + 02	4.91E + 01	5.10E + 01	2.12E + 01	5.08E + 02
	EA + EA	1.59E + 02	6.05E + 01	2.81E + 01	5.10E + 01	1.31E + 01	3.12E + 02
RI/ kg PM2.5 eq	GA + SMA	3.20E + 01	4.08E + 01	9.06E + 00	9.06E + 00	2.33E + 00	9.32E + 01
	EA + EA	3.00E + 01	1.27E + 01	5.18E + 00	9.06E + 00	1.81E + 00	5.87E + 01

3 Uncertainty Analysis

Uncertainty analysis in LCA includes data sensitive analysis, data uncertainty assessment and results uncertainty assessment. It is stipulated that uncertainty standard of SBDP is 10%. Firstly, sensitivity index is used to assess the influences of data quality to results reliability. Secondly, data uncertainty is performed aims at LCI data with sensitivity index larger than 10%. Thirdly, result uncertainty assessment of AI is performed. If result uncertainty indicator is larger than 10%, correct corresponding LCI data with data uncertainty indicator larger than 10%. If result uncertainty indicator is less than 10%, this demonstrates that the LCI data and results are credible. Sensitivity index is obtained from Eq. 2.

$$S_{mn} = (\Delta AI_m / AI_m) / (\Delta I_n / I_n) \quad (2)$$

where S_{mn} is sensitive index; AI_m is one analysis indicator; I_n is related LCI data of AI_m; ΔI_n is the variation of I_m; ΔAI_m is the variation of AI_m caused by ΔI_m. The higher the value of S_{mn}, the greater the influence of this data on AI.

LCI data could be classified as foreground process inventory data (obtained through research and calculation) and background process inventory data (obtained from existing database). The foreground and background process uncertainty indicator ($U_{foreground}$ and $U_{background}$) could be obtained through Eq. 3.

$$U_{foreground} = \sqrt{U_r^2 + U_i^2} = \sqrt{\sum_j^5 U_j^2 + U_i^2}\ U_{background} = \sqrt{U'^2 + U_3^2 + U_4^2 + U_5^2} \quad (3)$$

where *U'* is inventory uncertainty offered by selected database; U_r is foreground data uncertainty indicator; U_i is algorithm uncertainty indicator, 0 for data acquired directly, 0.025 for data after equalization, 0.050 for data obtained experientially, and 0.100 for data calculated theoretically; U_j is data uncertainty index values according to the Semi-quantitative Genealogical Matrix promoted by Ecoinvent database.

Monte Carlo Simulation was adopted to assess results uncertainty. The environmental impact factors are considered as random variables, and the analysis results are expressed as the result of statistical distribution (Eq. 4). This study assumes that the LCI data obey lognormal distribution.

$$\begin{aligned} AI &= g(X) \\ f(x) &= \frac{1}{x\sigma\sqrt{2\pi}} e^{-\frac{(\ln x-\mu)^2}{2\sigma^2}} \\ \sigma &= \sqrt{U_{foreground}^2 + U_{background}^2} \end{aligned} \quad (4)$$

where *X* is LCI data; σ is standard deviation. LCI data was sampled several times based on the probability distribution function. The analysis results are statistically analyzed and the results uncertainty indicator is determined according to their distribution parameters.

4 Results and Discussion

Uncertainty Analysis

The result uncertainty is shown in Table 4. Result uncertainty indicator of *AI* is less than 10%. Thus, it can be concluded that the output result of the SBDP environmental impact analysis model is credible.

Life Cycle Interpretation

Comparative Analysis of Environmental Impacts

Figure 2 shows the comparative analysis of *AI* of the two structures. It is noticed that "GA + SMA" structure contributes more environmental impacts than "EA + EA" structure. In the life cycle of the two typical paving layers, the stages with high energy consumption and emission are the raw material acquisition, following plant production and operation and maintenance.

Table 4 Result uncertainty indicator

AI	Structure	Full life cycle	Uncertainty indicator
PED/MJ	GA + SMA	3.25E + 06	7.32%
	EA + EA	2.46E + 06	6.46%
GWP/kg CO2 eq	GA + SMA	1.58E + 05	4.47%
	EA + EA	9.93E + 04	6.18%
AP/kg SO2 eq	GA + SMA	5.08E + 02	3.64%
	EA + EA	3.12E + 02	4.50%
RI/kg PM2.5 eq	GA + SMA	9.32E + 01	4.06%
	EA + EA	5.87E + 01	3.92%

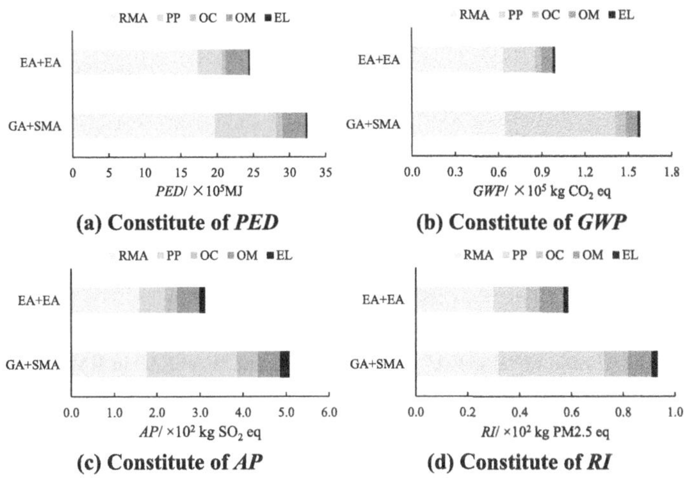

Fig. 2 Compare of *AI* of "EA + EA" and "GA + SMA"

The environmental impacts of the two pavement structures at different life cycle stages were analyzed (Fig. 3). The environmental impact of "EA + EA" and "GA + SMA" are similar during the stages of raw material acquisition, operation and maintenance, and end-of-life. However, in the stages of plant production and on-site construction, the environmental impact produced by "GA + SMA" is higher than "EA + EA". In the stage of plant production, compared with "EA + EA", "GA + SMA" consumes 2.5 times of energy and produces 3.4 times of gas approximately. In the stage of on-site construction, compared with "EA + EA", "GA + SMA" produces 1.7 times of environmental impacts approximately.

Key Factors of Environmental Impact

The comparison of environmental impacts at different life cycle stages of "EA + EA" structure is shown in Fig. 4a. Raw material acquisition stage takes 70.76% of total *PED* and contributes over 50% of total gas emissions. The focus of energy

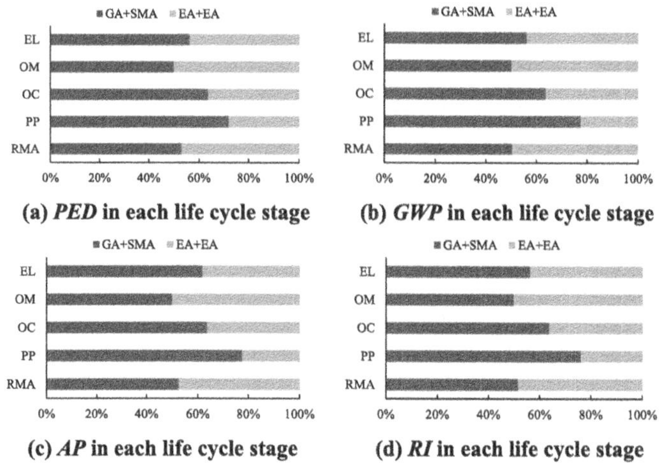

Fig. 3 Compare of *AI* in each life cycle stage

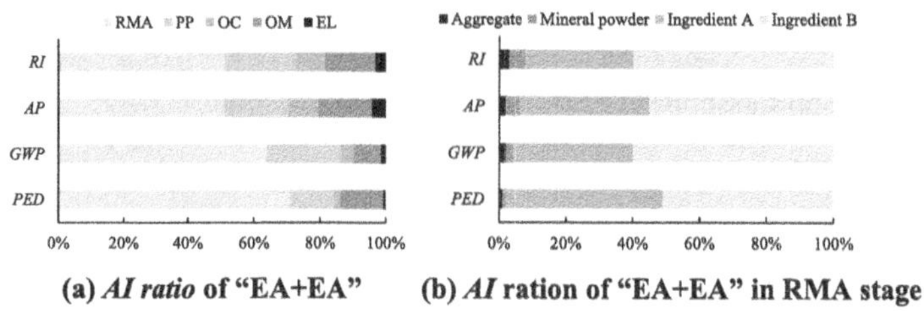

Fig. 4 Key factor analysis of "EA + EA"

conservation and emission reduction of "EA + EA" structure is on the stage of raw material acquisition. The environmental impact ratio of each continent of EA mixture is shown in Fig. 4b. It is considered that the most significant contribution to the environment impact is the production of EA part B, followed by the production of EA part A.

Comparison of environmental impacts at different life cycle stages of "GA + SMA" structure is shown in Fig. 5a. It indicates that most significant contribution to *PED* is raw material acquisition stage, and the main contribution to gas emissions is plant production stage. Figure 5b illustrate the comparison of *AI* during raw material acquisition stage between GA mixture and SMA mixture. The corresponding indicators of both materials are very close due to the considered data in LCI during related stages is quite similar. Plant production stage of "GA + SMA" structure produces most gas emissions, and the comparative contributions of both materials are shown in Fig. 5c. During plant production stage, mixing is the main process of gas emission, followed by drying process. Compared with SMA mixture, higher mixing temperatures and higher asphalt use make GA mixture consume more resources and discharge more emissions.

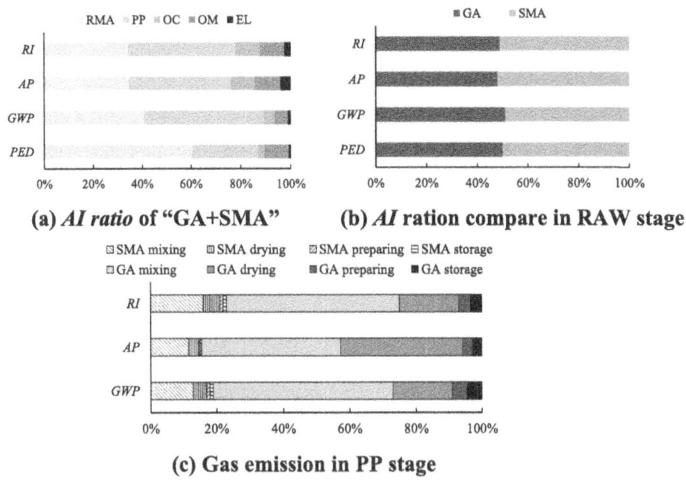

Fig. 5 Key factor analysis of "GA + SMA"

Overall, main problems during "GA + SMA" structure life time are the consumption of large amount of petroleum asphalt and high heating temperature of GA mixture. Moreover, emission reduction measures of plant product station are not applied efficiently because this stage produces over 40% of gas emission by consuming 26.17% energy. Therefore, optimizing the technology of producing petroleum asphalt binder and reducing the mixing temperature of GA mixture are the main means to reduce environmental impacts.

5 Conclusion

This study conducted the LCA for SBDP, including SMA, GA and EA mixtures. The conclusions are as follows:

1. Environmental impacts of "GA + SMA" is greater than "EA + EA". Raw material acquisition, plant production, operation and maintenance are the main stages that contribute to environmental impaction.
2. In the life cycle of "EA + EA" structure, the stage with the most significant contribution to the environment impact is the production of ingredient B of EA binder, followed by the production of ingredient A.
3. In the life cycle of "GA + SMA" structure, optimizing the technology of producing petroleum asphalt binder and reducing the mixing temperature of GA mixture are the main means to reduce environmental impacts.

References

1. Rebitzer G, Ekvall T, Frischknecht R et al (2004) Life cycle assessment: part 1: framework, goal and scope definition, inventory analysis, and applications. Environ Int 30(5):701–720
2. Leng Z, Al-Qadi IL, Cao R (2018) Life-cycle economic and environmental assessment of warm stone mastic asphalt. Transportmetrica a: Transp Sci 14(7):562–575
3. Wang T, Xiao F, Zhu X et al (2018) Energy consumption and environmental impact of rubberized asphalt pavement. J Clean Prod 180:139–158
4. Saeedzadeh R, Romanoschi SA, Akbariyeh N et al (2018) Sustainability assessment of recycled asphalt mixtures based on performance in full-scale testing. J Transp Eng Part B Pavements 144(2):04018024
5. Yang Y, Qian Z, Song X (2015) A pothole patching material for epoxy asphalt pavement on steel bridges: fatigue test and numerical analysis. Constr Build Mater 94:299–305
6. Chen L, Qian Z, Lu Q (2014) Crack initiation and propagation in epoxy asphalt concrete in the three-point bending test. Road Mater Pavement Des 15(3):507–520
7. Qian Z, Huang W (2014) Maintenance and Rehabilitation of Asphalt Pavement on Steel Bridge Deck. China Communications Press, Beijing
8. Ministry of Transport of P. R. China (2015). Specifications for Design of Highway Steel Bridge. China Communications Press, Beijing
9. Forster P, Ramaswamy V, Artaxo P et al (2007) Changes in atmospheric constituents and in radiative forcing. Chapter 2. Climate Change 2007. The Physical Science Basis
10. Jeroen G, Gorrée M, Heijungs R et al (2001) Handbook on life cycle assessment–operational guide to the ISO standards. Int J Life Cycle Assess 6(5):255
11. Jolliet O, Margni M, Charles R et al (2003) IMPACT 2002+: a new life cycle impact assessment methodology. Int J Life Cycle Assess 8:324–330
12. Swiss Center for Life Cycle Inventories (2013). Ecoinvent Database, Version 3.0. Saint Gallen, Switzerland
13. Blomberg T, Barnes J, Bernard F et al (2011) Life Cycle Inventory: Bitumen. European Bitumen Association, Brussels, Belgium
14. Zhenjiang environmental science institute (2014). Environmental impact statement of epoxy asphalt production plant project. Zhengjiang, China
15. European Commission (2010). ILCD Handbook: General Guide for Life Cycle Assessment—Detailed Guidance. Joint Research Centre, Belgium
16. Prowell B, Frank B, Osborne L et al (2014) Effects of WMA on plant energy and emissions and worker exposures to respirable fumes. National Cooperative Highway Research Program, Washington DC, USA
17. IKE, SCU-ISCP (2013). Chinese core Life Cycle Database version 0.8. IKE Environmental Technology Co., Ltd. & Institute for Sustainable Consumption and Production at Sichuan University
18. Ministry of Environment Protection of P. R. China (2014). Limits and measurement methods for exhaust pollutants from diesel engines of non-road mobile machinery (CHINA III, IV). China Environmental Science Press, Beijing
19. Huang Y, Spray A, Parry T (2013) Sensitivity analysis of methodological choices in road pavement LCA. Int J Life Cycle Assess 18(1):93–101

Bridge Extreme Event Loads: Earthquake, Wind and Fire

Efficient Fire Hazard Mitigation for Suspension Bridge Cables

J. Laigaard Jensen, N. Bitsch, and Harikrishna Narasimhan

Abstract Increasingly, large suspension bridges are exposed to fire risks as the traffic they carry increases. A truck fire on a suspension bridge may lead to main cable failure or a strength reduction that will either cause down-graded classification of the load carrying capacity or need for long lasting repairs of the main cable with large costs and traffic disruption as severe consequences. A serious fire in a truck occurred in 2013 on the New Little Belt Suspension Bridge in Denmark, and caused rapidly rising flame temperatures to above 1000 °C, somewhat similar to a hydrocarbon fire. Based on this accident and a number of other incidents with fires on roads and bridges, a risk and cost–benefit study focusing on fire risks for this bridge was carried out. Based on the result of the study, it was decided to provide fire protection to the main cables on the New Little Belt Bridge. Due to the large socio economic importance, the need for similar mitigation of fire hazards has been considered for bridges such as Älvsborg Suspension Bridge (Sweden), Great Belt East Bridge (Denmark) and A.L. MacDonald Suspension Bridge (Canada). Current status is that fire retro protection projects have been decided and are being prepared for the latter two bridges. Based on the accident on New Little Belt Bridge, a fire protection concept for main cables has been developed after detailed studies of the fire impact on this occasion. This has resulted in advanced modelling of the fire loading on a main cable and in design elaboration of an efficient fire hazard mitigation concept for main cables. This paper describes the systematic process for evaluation of the fire accident on New Little Belt Bridge and how this has evolved into a very efficient fire hazard mitigation concept for main cables. The concept may be used on existing as well as on new suspension bridges being essential elements in national road infrastructure systems.

1 Introduction

Considerations of protecting suspension bridge cables from fire are not yet standard design practice and not addressed in codes and design guidelines as it is for bridge

---

J. L. Jensen (✉) · N. Bitsch · H. Narasimhan
COWI A/S, Kongens Lyngby, Denmark
e-mail: lai@cowi.com

P. Gülkan et al. (eds.), *Developments in International Bridge Engineering*, Springer Tracts on Transportation and Traffic 17, https://doi.org/10.1007/978-3-030-59169-4_20

Fig. 1 Truck vehicle fire at New Little Belt Bridge at central node of main cable

stay cables at least in some places of the world [1]. However large suspension bridges are most often vital to society, and they are increasingly exposed to the risk of serious truck or tanker fires as the traffic carried by them increases. Experience shows that even a fire in a truck without dangerous goods, like the serious fire in 2013 [2, 3] on the New Little Belt Bridge (from 1970), may cause rapidly rising flame temperatures to above 1000 °C (Fig. 1). Hence, this is equivalent to the typical temperatures in hydrocarbon fires though in the truck fire case with a maximum heat release smaller than the actual hydrocarbon fire. The relatively high traffic frequency of such trucks compared to fuel or gas tankers, means that such truck fires are much more likely to happen.

With these high temperatures, a truck fire accident on the bridge may lead to cable failure or a strength reduction that will either cause down classification of the load carrying capacity or need for extensive and long lasting repairs or replacement of the main cable. Such outcomes will have high costs for the bridge owner as well as society. Thus, there is a large need for assessing the fire risk of such bridge cable systems and for developing adequate fire protection concepts.

2 Risk Assessment

For the New Little Belt Bridge, no comprehensive Operational Risk Assessment (ORA) was available to assess the magnitude of fire risks to the bridge and its acceptability. Hence, after the 2013 fire accident, the owner decided to assess the fire risk for the main cables due to the traffic by a cost–benefit based risk study.

The assessment concluded that due to the low traffic of fuel and gas tankers, only the risk from fire in trucks without dangerous (but still flammable) goods was significant with respect to the estimated frequency. As an example, serious consequences of such an accident could occur if a release of fuel from the diesel tank in a truck collision spread along the length of the truck and was ignited by hot mechanical parts or continuous sparks. Subsequently, the entire load on the truck could soon be on fire. Both modern packaging and materials in goods have become more and more flammable over the last decades. This means that a truck fires may expose bridges with almost the same high temperatures as hydrocarbon fires.

Due to the experienced high temperatures in the accident on Little Belt and previous truck accidents, as well as the results of tunnel truck fire experiences, it was concluded in the risk study that the fire temperatures should be modelled as that in a hydrocarbon pool fire at the road deck level equivalent to the fire code of Eurocode EN 1991-1-2 hydrocarbon time–temperature with maximum of 1100 °C. However, it was also noted that the fire modelling should take into account the geometry (relatively narrow width compared to length) of the truck to obtain the correct heights of the continuous flame (region with constantly high temperatures above 1000 °C) as well as the intermittent flame (region with temperatures decreasing while moving up the flame/plume and the flame fluctuating).

3 Modelling of Heat Penetration

A large effort was given to the modelling of the heat transfer into cable as basis for designing the required fire protection system. Based on the results of the investigation of the fire damages at New Little Belt Bridge and a study of sparse but relevant literature, a detailed process of modelling was carried out for the risk study and subsequent fire protection project.

The reason for this was that the prediction of heat penetration into structures is complicated as it normally involves mechanisms as conduction, radiation and sometimes also convection. For protection of cable systems, the heat penetration complicates further as cables contains a significant amount of air voids and at the same time interaction with other materials than steel, e.g. zinc coating elastomer wrapping. Hence, a large effort has to be made to model the heat transfer to the cable cross section in question although on the other hand a long cable can generally be described by a 2-dimensional axial-symmetric model, which provide some simplification.

For the New Little Belt, the heat penetration into the cable was modelled in two different ways. One was based on a model from NIST [4], which results in a rather quick heat-up of the cable all through the wires. This is a method more suitable for cables with parallel wires where the wire-to-wire contact is based on parallel aligned wires. However, the New Little Belt Bridge cables are made from 61 helical strands with 9,000 wires being not parallel but rather having a wire-to-wire point contact. The void percentage is thus rather high at about 35%. This means that the outer strand layer will be expected to be heated more in the start of the fire than predicted by the NIST model. Hence, a second COWI in-house model was therefore developed for the cross-sectional heating transfer properties to consider this effect and this model was used in determination of how well the cable should be protected to prevent any significant loss of load bearing capacity in any of the cable wires during the design fire.

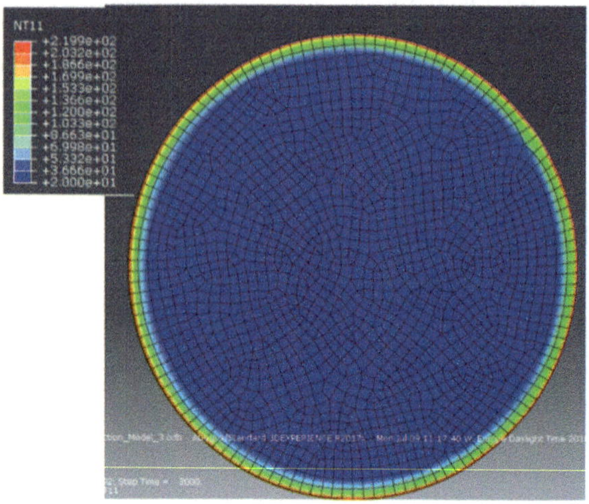

Fig. 2 ABAQUS modelling was used for the deriving the model of the heat transfer

As the algorithm of the heat transfer is quite complicated, this was later in connection with the fire protection of the Great Belt Bridge (Sect. 6) verified by an ABAQUS model as shown in Fig. 2, which also was used for predicting heat penetration into the cable when combined solid structural parts of the cable system. The agreement between the models was better than 4% and the ABAQUS model was also checked against a test (Fig. 5) to be within 25% accuracy.

4 Passive Fire Protection of Main Cable

A passive fire protection system was developed for the main cable with installation of a fire insulation protection mat around the cable and with protection of a stainless steel cap. Figure 3 shows the concept with photos from the site works.

With the developed models, the passive fire protection for the main cable was based on the following main principles and as summarized in Table 1:

- The height of fire protection was determined to be 10 m above deck level based on a cost–benefit study. The redundancy of hangers meant that it was sufficient to insulate the main cable.
- The fire load was defined by the fire code of Eurocode EN 1991-1-2 hydrocarbon time–temperature with maximum of 1100 °C. A 30 min duration was defined equivalent to the effective response time of fire brigade task force.
- Design criteria for the fire protection was that it should protect the cable steel against temperatures above 400 °C. Furnace testing of some of the original cold drawn cable wires showed that no irreversible loss of tensile strength was found after cooling down from a temperature of 400 °C.

Fig. 3 Fire protection works for main cable. Left: Stainless steel cap is being installed on top of fire protection mat. Right: Works at the central node in the main span

Table 1 Technical requirements for the passive fire protection

Assumptions	Values
Fire temperature on the steel cap	1100 °C
Duration of hydrocarbon fire, min	½ h
Max. temperature of cable wires	400 °C
Height of fire protection above deck level	10 m
PFP material	
Max. thickness	28 mm
No. of layers, min., offset	2
Max. thermal conductivity, thickness 10 mm	0.075 W/m/K
Max. thermal conductivity, thickness 15 mm	0.115 W/M/K
Max. thermal conductivity, thickness 20 mm	0.150 W/M/K
Max. thermal conductivity, thickness 25 mm	0.190 W/M/K
Min. compressive strength	0.1 MPa
Water absorption, max	1% vol
Weight, max. per m cable	8 kg
Length of insulated cable	624 m

- A removable stainless steel cap should protect the passive fire protection (PFP) against weather and mechanical damages.
- The design should allow for a main cable inspection car to run on the steel cap without damaging the PFP.
- The concept should not allow any accumulation of water that could cause corrosion to the cable and the steel cap.

- The cable clamps should also be fire protected as these do not protect the main cable against fire.

5 Method of Installation

Based on the above assessment, a concept design for application of fire protection was developed for the New Little Belt Bridge in the spring of 2017. Following a tendering process, the fire protection was installed in the autumn of 2017. Figure 3 and 4 show photos from the installation works.

The concept design developed comprised a fire insulation mat with the required properties, 15 mm in two offset layers and protected by a stainless steel cap fixed by steel straps to ensure that the concept would maintain its integrity over the duration of a fire.

The concept design was based on prefabricated segments 1 m long with special transition segments at cable clamps and sleeves of the main cable dehumidification system. The objective of the dehumidification system is to protect the main cable against corrosion.

Specific requirements to the concept design were that:

- The concept design shall be able to carry the load from a main cable inspection car corresponding to a wheel pressure of 0.1 MPa without damaging the PFP.
- The PFP material shall be hydrophobic to avoid accumulation of water around the cable.
- The concept design shall allow for adequate drainage to avoid accumulation of water.
- The concept shall be easy to install and remove so that the amount of traffic restrictions would be as little as possible and of short duration, both during installation and at inspections of the main cable every five years. The latter is expected to be required as the main cable is dehumidified and the condition as well as the air

Fig. 4 Left: Installed stainless steel cap with walking surface. Right: Adaption section at a dehumidification sleeve

Fig. 5 Left: Test model. Right: Test model in furnace chamber after a test completion

tightness of the cable must be checked at regular intervals for the dehumidification to provide effective corrosion protection to the main cable.

As part of the design phase and prior to tendering and installation, a prototype of the PFP was developed and a trial assembly was carried out on the bridge to ensure that it would fit and had sufficient tolerances to account for variations in the main cable geometry. Furthermore, the trial assembly also demonstrated that it would be possible to install segments in a time-efficient manner by use of lift.

The installation work was carried out with only a closure of the emergency lane and use of low scaffolding and a lift for heights up to 10 m above the roadway. In total, about 624 m of main cable were protected by this PFP concept. Each main cable of the bridge is about 1500 m long.

6 Great Belt East Bridge Fire Protection Project

After the fire protection of the New Little Belt project, a similar fire protection has been prepared and tendered in 2018 for the Great Belt East Bridge. Based on the experiences from the fire on New Little Belt and the fact that vehicle fires occur quite frequently, it was decided also to protect the main cables of the Great Belt Bridge based with a concept quite similar to the one successfully installed on the New Little Belt Bridge. The length of each main cable of the Great Belt East Bridge is approximately 3 km each, which is double the main cable length of the New Little Belt Bridge.

Also in this case, the original type of cold drawn cable steel wire was tested for possible permanent capacity loss after being heated to 400 °C. The testing confirmed this threshold. It may be added that at about 420 °C, the heat will lead to the melting of the zinc on the galvanized cable wires, which therefore gave a secondary threshold temperature criteria not to be exceeded.

However to optimize and quality the concept further for the Great Belt Bridge, it was decided to carry out testing of 1:1 modelling of 5 m long fire protection segment in a fire laboratory furnace. The purpose was to investigate the efficiency of different fire isolation materials, the overall integrity of the concept and also the procedure of installation of the fire protection. Figure 5 shows the test model. The results confirmed the effectiveness and integrity of the developed fire protection concept. Measured steel temperatures were well below 400 °C.

The Great Belt East Bridge main cable is designed with parallel wires giving a void ratio of about 21%, i.e. much lower than on the New Little Belt Bridge (where it is about 35%). With more than 18,000 wires, it was considered not realistic to make a fire test on a real cable; a steel pipe is thus used to test the insulation concept in a 30 min fire. The test on the pipe provides a good basis for obtaining a reasonably accurate value for the thermal conductivity of the tested materials at very high temperatures close to the maximum flame temperatures of about 1100 °C.

As for New Little Belt Bridge, one important aspect to designing an adequate fire protection for the Great Belt Bridge was to model the heat transfer in the cable system correctly. Therefore again large emphasis was given to this modelling approach and comparison and calibration with the test results were also carried out.

The NIST method [4] for calculating the temperature build-up into the steel cable is a conservative approach with respect to the rupture and loss of strength scenario, i.e. giving a faster heat penetration to the inner parts of the cable. Because it is desired that a significant loss of capacity and also damage of all wires including also the outer wires shall be prevented by the PFP, a model of the same type as on the New Little Belt Bridge, i.e. with a relatively low heat penetration, is used to specify the thickness of the PFP material. The combined use of the NIST approach to establish criteria for capacity loss and the New Little Bridge approach to establish criteria for temperature exceedance for the main cable outer wires thereby provides a reasonably robust basis for establishing the PFP requirements for the Great Belt East Bridge.

7 Protection of Steel Works by Intumescent Coating

A remaining fire risk mitigation to be carried out is to apply a PFP, probably in the form of a thick film intumescent coating applied to cable clamps. To provide sufficient protection of the main cable, the clamps will have to be protected to the same height. The installation of this coating has not yet been carried out. Figure 6 shows an example of the heat transfer model of the interface between cable clamp and cable.

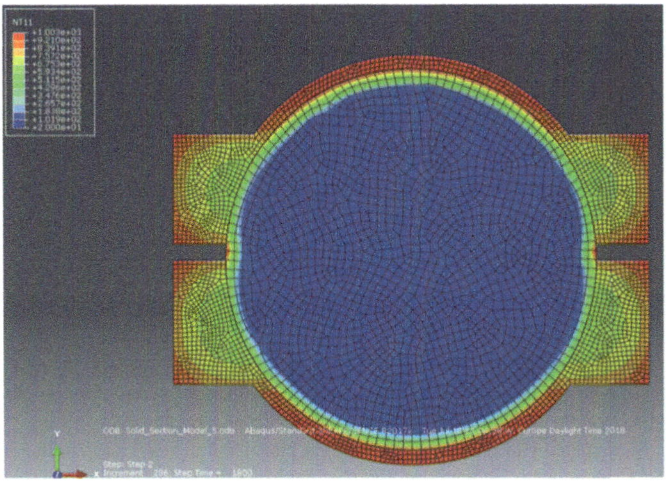

Fig. 6 Example from Great Belt Bridge with ABAQUS modelling of unprotected cable clamp with temperatures exceeding 400 °C

Meanwhile COWI has just started, with funding from the trust fund COWIfonden, a one year research project on the use of intumescent coating for protection of bridge cable systems in corporation with academia (DTU and UoE). The project will based on large scale testing look into important issues such as:

- The insulation properties of the coating as function of temperature.
- The mechanical robustness of coating together with its durability with respect to the quite harsh ambient offshore environment.
- Construction requirements for steel surface preparation and steel works.
- A modelling approach for defining the required extent of the isolation and the influence of heat transfer from unprotected parts of steel work.

When these issues have been solved, it is expected that the combination of insulation mats together with intumescent coating will provide a basis for flexible and robust approach on the fire protection of bridge cable systems.

8 Conclusion

The developed passive fire protection concept for the main cables of large suspension bridges has been developed ensuring adequate robustness against the most likely fires on the bridge. The design objective has not only been to avoid a bridge collapse but to avoid loss of bearing capacity as the replacement of a suspension main cable is very costly. The adequateness of the concept has been demonstrated by extensive modelling as well as testing.

Further work will be carried out to improve heat transfer modelling in cables further just as the concept is planned to be extended with use of intumescent coating to obtain a more flexible concept with can be tailor made to the individual bridge and the particular needs of the owner. The intention is also that the flexibility of the concept will allow the concept to be used efficiently for new bridges as well as for retrofit of existing bridges.

References

1. Recommendations for Stay Cable Design (2012). Testing, and Installation, PTI DC45.1–12
2. Bitsch N, Larsen J, Stoklund E et al (2014) Fire on the New Little Belt Bridge. In: IABSE Madrid Symposium: Engineering for Progress, Nature and People, pp 2729–2736
3. Kragh E, Narasimhan H, Laigaard JJ (2018) Fire protection of suspension bridge main cables. In: IABSE Conference – Engineering the Past, to Meet the Needs of the Future, 25–27 June 2018, Copenhagen, Denmark
4. Safety Assessment of Parallel Wire Suspension Bridge Cables Under Thermal Effects (2010). NIST Technical Note 1678

Seismic Performance of Bridge Systems Enhanced with Cellular-Solid Shear Walls

Spyridoula M. Papathanasiou, Panos Tsopelas, and Thanasis Zisis

Abstract Light-weight shear wall panels with deterministic cellular periodic architecture could provide the basis for vibration mitigation in large scale structural systems. In this study the behavior of shear walls with cellular solids under seismic loading is examined. An investigation to evaluate the effects of different cellular configurations is conducted, including the orientation angles and the shapes of the cells (honeycomb, re-entrant or chiral architecture). The cellular walls may be appropriately arranged between the columns of a pier of a concrete bridge system for seismic resistance enhancement in both directions. Appropriate simplified Finite Element Models are developed to predict the stiffness, strength, and energy dissipation effectiveness of shear wall panels when subjected to monotonic and cyclic shear loading, with ABAQUS software. The total column-cellular wall-deck system behavior is studied with simple stick models after the calibrated wall properties have been taken into account.

1 Introduction

Cellular solids are assemblies that result from appropriate interconnection of solid struts or plates [1]. The honeycomb configuration has been widely used as the filling material of sandwich composite panels, while it has also been implemented in other fields such as aerodynamics for controlling the wind turbulence, in automotive crash test barriers for absorbing energy [2] and in thermal insulator materials due to their low thermal conductivity. Many natural materials appear having similar constructions as the man-made cellular solids: woods, plant leaves, trabecular bones or fruits. The ultra-low-density format, the light weight and the good thermal behavior have resulted in cellular solids to receive great attention in developing innovative structural systems with enhanced stiffness and strength. A light-weight steel shear wall panel of deterministic-cellular-periodic-architecture could provide the basis for an efficient

S. M. Papathanasiou (✉) · P. Tsopelas · T. Zisis
Department of Mechanics, School of Applied Mathematical and Physical Sciences, National Technical University of Athens (NTUA), Athens, Greece
e-mail: spapath@central.ntua.gr

P. Gülkan et al. (eds.), *Developments in International Bridge Engineering*, Springer Tracts on Transportation and Traffic 17, https://doi.org/10.1007/978-3-030-59169-4_21

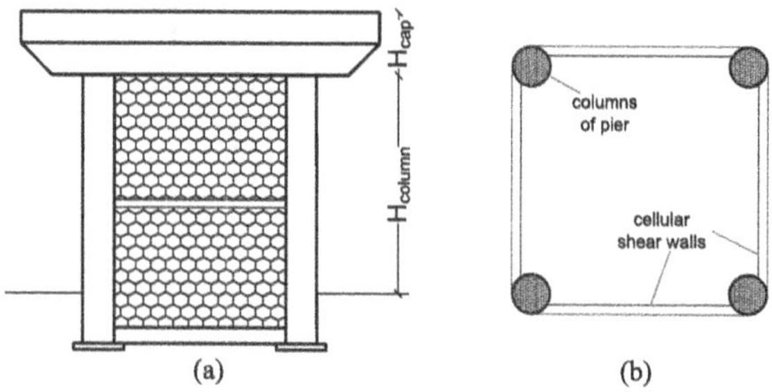

Fig. 1 **a** Cellular shear wall (honeycomb) arranged between double-column pier of a bridge, **b** Plan view of a four-column pier with cellular walls applied in both directions (Keller and Bruneau [4]).

and cost effective response modification system for seismic vibration mitigation in large scale structural systems.

The idea of using shear wall panels with cellular configurations is not new though. Vian and Bruneau (2005) [3] proposed the use of shear wall panels with perforations to alleviate the over-strength concerns of solid steel plate shear walls. In the limit, when the number of perforation is maximized and the size of perforation is minimized, a steel shear wall panel with perforations turns to a cellular-solid shear wall panel. Keller and Bruneau [4] presented some interesting results of the analysis performed when steel plate shear walls, both perforated and not, were attached between the double and triple column piers of bridge models. Tsopelas and Chen [5] studied the seismic behavior of cellular-solid-shear walls with different cell orientation. A perforated-aluminum-shear panel in combination with steel braces was also proposed by Foti et al. in 2009 as an energy dissipation mechanism of a frame [6].

The main objective of this work is to evaluate the effects of different cellular configurations, including the orientation angles and the shapes of the cells (honeycomb, reentrant or chiral architecture), under seismic loading. Appropriate simplified Finite Element Models are developed to predict the stiffness, strength, and energy dissipation effectiveness of shear wall panels, with ABAQUS software, when subjected to monotonic and cyclic shear loading. Further, the different cellular wall configurations are appropriately arranged between the double-column pier (Fig. 1a) of a concrete bridge system for a transverse seismic resistance enhancement. This can apply in both directions in case the pier consists of four columns (Fig. 1b) for strengthening in both directions. The combined column-cellular wall-deck system is studied with simple stick models after the calibrated properties of the different shear wall configurations have been taken into account.

2 Cellular-Solid Shear Wall Systems

Cellular-solid shear wall systems outweigh the classic steel shear wall panels due to: a) being less prone to out of plane bucking, b) weighing less, c) offering flexibility to the structure attached to, while dissipating energy through the plastic strains they suffer and d) offering architectural elegance. The studying of their behavior to certain load cases has revealed that the geometry of the basic cell, the so-called "unit cell", affects a lot their mechanical properties. In the ideal case when it is equiaxed, the cell can be considered isotropic. In case the unit cell is slightly elongated or flattened, the properties of it appear to differ significantly between the two vertical directions x and y. In reality the mechanical properties of unit cells are neither isotropic nor homogeneous on their microstructure. Only the macroscopic behavior of cellular solids is considered homogeneous, given that the cell size is relatively small compared to the size of the structural element consisting of cellular solid.

Three different cellular configurations are chosen to be studied, in order to reveal the sensitivity of the overall shear behavior to the shapes and orientations of unit cells with respect to an initial frame. Figure (2a) depicts a honeycomb cellular shear wall, while Fig. (2b) shows three different orientations for the honeycomb unit cell to be compared. The effect on the strength and energy dissipation properties of the shear walls is quantified by analyzing detailed Finite Element Models with ABAQUS software.

The honeycomb assembly chosen for the Finite Element Analyses is not the regular one ($h = l = 1$) but is slightly elongated, with the ratio of its wall lengths to be $h/l \approx 1.1$. Two more configurations were utilized for the unit shells of shear walls, in compliance with honeycomb's geometry: the reentrant and the chiral unit cells (Fig. 3) with the same variation of orientation angles. Both honeycomb and reentrant geometries have the same ratio $l/t \approx 4$. Especially for the chiral unit cell the topology

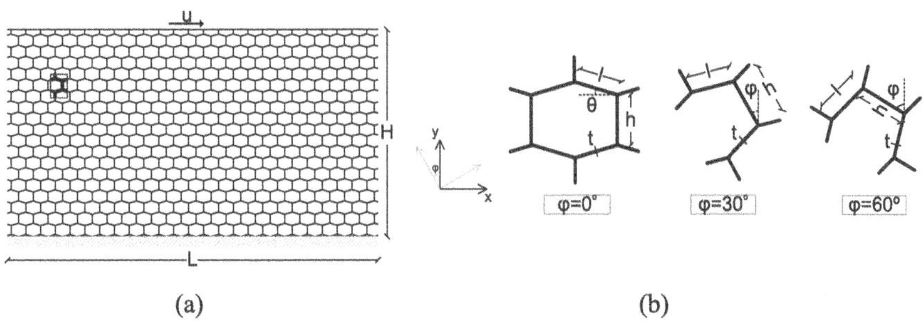

Fig. 2 **a** Honeycomb Cellular Shear Wall of dimensions L × H, **b** Different orientations ($\varphi = 0^{\circ}$, 30°, 60°) for honeycomb unit cells used in shear walls of this study.

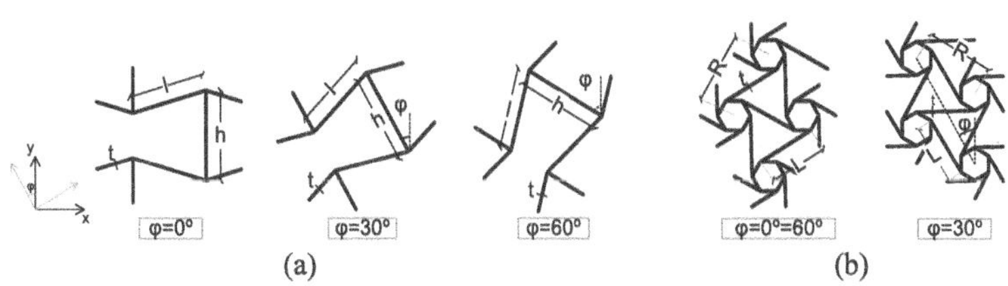

Fig. 3 **a** Different orientations ($\varphi = 0°, 30°, 60°$) for reentrant unit cells and, **b** chiral unit cells used in shear walls of this study.

aspect ratio [7] was depicted so as to be L/R = 0.87. For comparison purposes the thickness of the cell walls was chosen so as to correspond to the same relative density value $\rho^* = 0.3$ ($=\rho_{cell}/\rho_{solid}$) for each unit cell.

2.1 Modeling of Cellular Solid

Simple beam elements have been successfully used by many researchers for modeling both the linear and nonlinear mechanical properties of cellular materials (Papka and Kyriakides 1994, Overaker et al. 1998) [8, 9]. In this work, the ABAQUS [10] software is used for the analyses. The element B21 is used for modeling the walls of the cells, while appropriate meshing was chosen so as both bending and shearing effects to be monitored. The overall dimensions of the walls are H × L × d (d is the depth of a cell and of the wall).

A parametric study is conducted firstly to quantify the mechanical properties of the cellular solid shear wall panels under monotonic shear loading as a function of the shape of unit cell used (honeycomb or reentrant or chiral) and the vertical cell wall orientation angle (φ), for a material with given yield strength. The shear loading is applied as lateral deformation (u) of increasing amplitude on the top of the shear wall considered (see Fig. 2a). The ratio of the applied deformation at the top of the wall, u, over its height, H, is defined as average shear strain ($\gamma_{aver} = u/H$). Similarly, the reaction force at the bottom of the wall normalized by the cross sectional area of the shear wall panel is defined as the average shear stress of the cellular solid ($\tau_{aver} = F/d/L$). This normalization of the lateral displacements and reaction forces makes the results scalable under the assumptions of the analysis and aids the development of "spring" like hysteretic models for full scale cellular shear walls. The material considered is high yield steel with properties shown in Table 1. A multi-linear kinematic hardening material model was utilized in the analyses performed. The dimensions of the shear wall were chosen so as H/L = 0.54 (=4.2/8.8 m).

The effect of the cell shape and orientation on the mechanical properties of cellular solids is demonstrated by plots of τ_{aver} vs γ_{aver} in Fig. 4, for monotonic loading in both directions. Large discrepancies in stiffness, strength and plastic strains are observed for the different cellular assemblies of the same wall. The behavior of the same cellular solid wall (H/L = 4.2 m/8.8 m = 0.54) consisting of three different unit cell shapes, oriented by $\varphi = 0°$ and $\varphi = 60°$ is studied under cyclic loading in Fig. 5 and the hysteretic responses are symmetric. This is due to absence of buckling of cells of the wall modeled. When the cell walls are under compression they may suffer severe buckling and this may lead to an important load resistance reduction. The resulting

Table 1 Properties of steel material

	Young's modulus, E (GPa)	Yielding strength, σ_y (MPa)	Poisson's ratio, ν
Steel	200	450	0.30

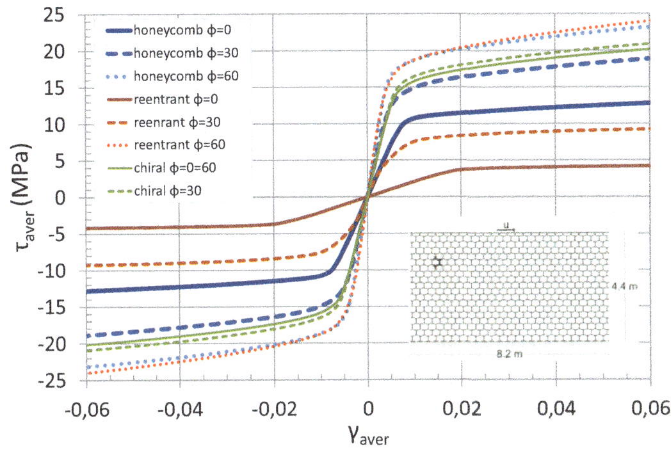

Fig. 4 Effect of the shape and orientation of unit cell on the τ_{aver} vs γ_{aver} response of a cellular solid wall in monotonic loading

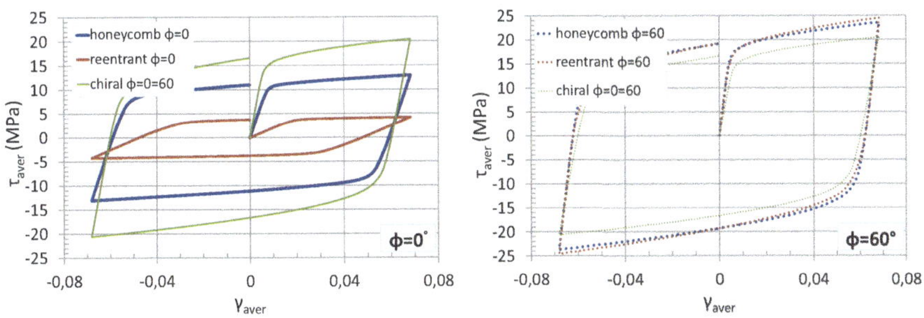

(a) (b)

Fig. 5 **a** Effect of the shape of unit cell on the τ_{aver} vs γ_{aver} response of a cellular solid wall in cyclic loading for orientation angle $\varphi = 0°$ and **b** $\varphi = 60°$.

symmetric hysteretic behavior outweighs the response of a brace in a braced-frame structure under cyclic loading, which is more likely to experience buckling.

According to Fig. 5 the cell walls appear to yield almost uniformly resulting in stable hysteretic response with significant yielding strength and post yielding stiffness for the case of chiral geometry ($\varphi = 0° = 60°$, see Fig. 3b.) On the other hand, the same wall consisting of reentrant cells of same relative density and same orientation ($\varphi = 0°$) appear to develop much smaller elastic stiffness, almost zero post yielding stiffness and overall dissipate much lesser incoming energy. The honeycomb geometry lies between the two aformentioned behaviors. For the case that $\varphi = 60°$, all three geometries appear to have almost identical properties and dissipate important amount of seismic energy.

3 Response of a Double-Column Bridge System Fitted with Cellular-Solid Shear Wall Under Seismic Excitation

To determine the effect of a cellular shear wall on the overall seismic behavior of a bridge pier we consider a double-column pier model of a concrete bridge inspired by the one proposed by Keller and Bruneau [4], referred as Prototype Bridge #2. The original bridge is a three span one, with two piers of 4 columns each, two at each direction. For simplicity we consider the equivalent double column pier (transverse direction of the original bridge) that consists of one cellular wall along the total height that is made of four identical cellular shear panels. A brief sketch of the bridge with the four shear panels of the wall is given in Fig. 6. Each panel is connected with the bridge only at its top and bottom faces and that is the reason why horizontal steel strips are considered between them. The properties of the bridge and the cellular shear panels and wall are given in Table 2. The masses of the walls are neglected. The values of the properties agree with the proposed ones by Keller and Bruneau [4] after being appropriately modified for the case only two columns are considered. The stiffness of the system in the transverse direction and the percentage of participation of each part to the overall stiffness [shear wall: 32%, concrete frame: 68%] are

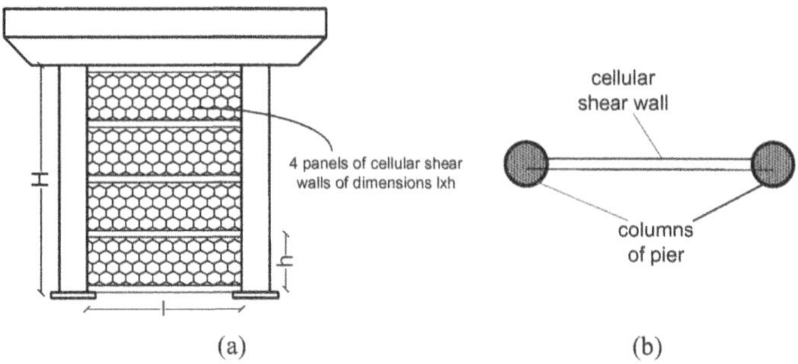

Fig. 6 **a** The two column pier considered with 4 cellular panels attached at their top and bottom faces. **b** Plan view of the double-column pier with the wall applied in transverse direction.

Table 2 Bridge and cellular walls properties

Total height of two columns, H (m)	9.37
Total length between two columns L = l (m)	3.71
Dimensions of cellular panels of the wall, l × h (m)	3.71 × 2.34
Mass of superstructure, pier cap and columns, m_{tot} (Mg or tons)	1231.35
Stiffness of each elastic concrete column, k_{col} (MN/m)	34.00
Poisson's ratio of both concrete and steel, ν	0.30

identical to the one presented by the authors after pushover analysis was performed in the same report [4].

The two concrete columns are considered to have elastic behavior. In order to use the results of Fig. 4 that correspond to the shear behavior of a wall of dimensions 8.2 × 4.4, the ratio of the total dimensions l/h is examined, based on the fact that the effective shear modulus of one wall will be identical to any other with the same ratio l/h, made of the same material. For simplicity we consider that the ratios l/h of the wall already analyzed and each of the four panels applied between the pier columns are almost the same (8.2/4.4≈3.71/2.34). The depth of the cellular wall was picked d = 0.06 m so as the honeycomb configuration of orientation $\varphi = 0°$ to correspond to the desired stiffness compared to the overall one ($K_{wall} = 0.32K_{total}$), as described before. The same depth was considered for all the other configurations as well. The shear modulus G and the corresponding stiffness of each cellular panel configuration are presented in Table 3.

The total stiffness of the wall according to the stiffnesses of each of the four panels (Table 3) can be estimated if the panels are considered as springs connected in series. The total wall's and pier's properties are presented in Table 4. The pier is analyzed with a simple stick model, considering that the total mass of the pier is concentrated at the top of the columns height and that the overall stiffness is elastic–plastic according to the findings of Table 4. The pier's damping ratio was considered $\xi = 2\%$.

One set of twenty ground motion time histories was used as vibration of the stick model, identical to the far field motions used by Tsopelas et al. [11]. The motions are appropriately scaled so as the average spectrum to match the target design spectrum as presented in AASHTO for soil type II, A = 0.4. The results of time history "Eq01" (Landers seismic event 1992) are presented in Fig. 7 in terms of total pier force vs deck displacement (relative to the ground) for the three cellular geometries compared before in Fig. (5a): regular honeycomb, reentrant honeycomb and chiral for $\varphi = 0°$.

Table 3 Properties of cellular panels for the different configurations (Fig. 4)

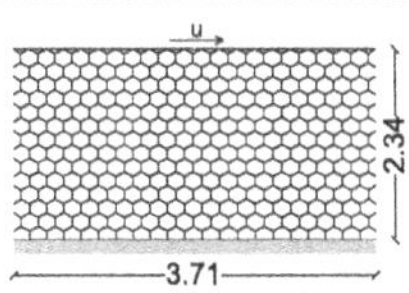

	Shear Modulus, $G=t_{aver}/\gamma_{aver}$ (MN/m^2)		**Stiffness for t=0.06m, K=GA/h (MN/m)**		**Yield Force (MN)**
Cellular Pattern of panel	**Elastic, G_{el}**	**Plastic, G_{pl}**	**Elastic, K_{el}**	**Plastic, K_{pl}**	**F_y**
Honeycomb φ=0°	1342.00	48.80	127.66	4.64	2.30
Honeycomb φ=30°	2180.00	101.80	207.38	9.68	3.05
Honeycomb φ=60°	3488.00	109.25	331.81	10.39	3.89
Reentrant φ=0°	200.00	15.69	19.025	1.49	0.79
Reentrant φ=30°	1268.00	34.45	120.62	3.28	1.66
Reentrant φ=60°	3725.00	136.88	354.35	13.02	3.72
Chiral φ=0°and 60°	2285.00	130.70	217.37	12.43	3.29
Chiral φ=30°	2670.00	108.72	254.00	10.34	3.39

Table 4 Properties of cellular wall for the different configurations

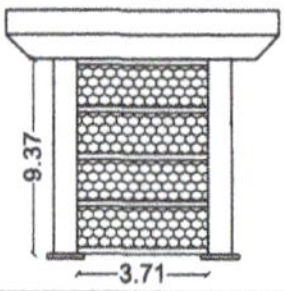

Cellular pattern of panel	Stiffness of total wall for t=0.06m, K=GA/h (MN/m)		Stiffness of pier= stiffness of total wall + 2*stiffness of column (MN/m)		Yield Force (MN)
	Elastic, $K_{wall,\ el}$	Plastic, $K_{wall,\ pl}$	Elastic, $K_{pier,\ el}$	Plastic, $K_{pier,\ pl}$	$F_{pier,\ y}$
Honeycomb φ=0°	31.92	1.16	99.92	69.16	7.20
Honeycomb φ=30°	51.84	2.42	119.85	70.42	7.06
Honeycomb φ=60°	82.95	2.60	150.95	70.60	7.09
Reentrant φ=0°	4.76	0.37	72.76	68.37	12.09
Reentrant φ=30°	30.16	0.82	98.16	68.82	5.42
Reentrant φ=60°	88.59	3.26	156.59	71.26	6.57
Chiral φ=0°and 60°	54.34	3.11	122.34	71.11	7.40
Chiral φ=30°	63.50	2.59	131 50	70 58	7.03

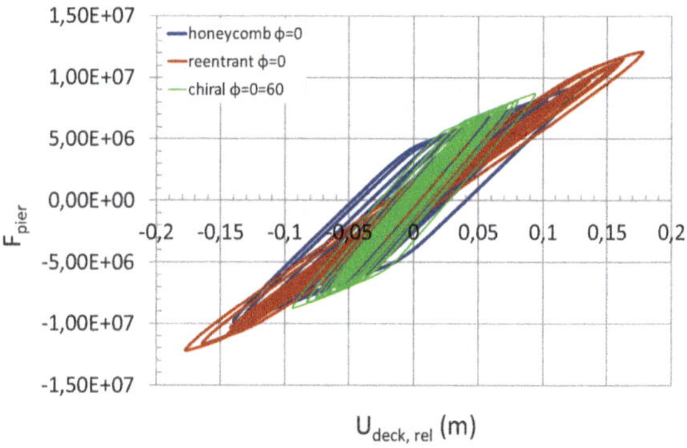

Fig. 7 Total force of the pier vs relative displacement of the deck for three different cellular wall geometries with orientation $\varphi = 0°$: regular honeycomb, reentrant honeycomb and chiral

For the specific seismic event, the pier with the regular honeycomb pattern of the cellular wall appears to dissipate the larger amount of incoming energy compared to the others. The pier with the chiral cellular wall corresponds to smaller amount of dissipated energy but to greater both elastic and post yielding stiffness. Finally the pier with the reentrant wall exhibits elastic behavior due to the maximum displacement imposed by “Eq01” seismic event being smaller than its yield displacement.

According to the results in Fig. (8), in case that the three wall geometries (honeycomb-reentrant-chiral) are oriented at $\varphi = 60°$ the overall pier behavior changes. As shown in Fig. (4) for angle $\varphi = 60°$ the wall panels show the largest elastic and post-yielding stiffnesses of all the other orientations. In addition, all three wall configurations dissipate more energy compared to the orientation of $\varphi = 0°$.

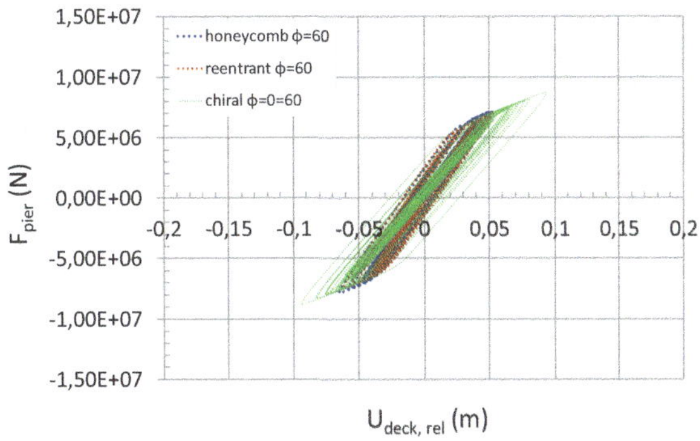

Fig. 8 Total force of the pier vs relative displacement of the deck for three different cellular wall geometries with orientation $\varphi = 60°$: regular honeycomb, reentrant honeycomb and chiral

Obviously, this is translated to much smaller lateral displacements but not necessarily significantly different pier forces.

4 Conclusions

According to the present study, cellular shear walls can be used as an effective response modification assemblies for seismic vibration mitigation in large scale structural systems. The most important conclusions are: a) the cellular shear walls are lightweight structural members which may easily fit between the columns of a pier and are not prone to severe buckling effects and b) depending on the geometry and orientation of unit cells chosen, the walls can be designed to exhibit very different behavior: they may dissipate large amounts of energy through plastic displacements or they may contribute additional stiffness to the overall pier system and make it experience very small displacements. The analyses showed that the same pier exposed to a specific seismic event appears to have a range of different responses, from elastic to fully plastic, depending on the unit cell shape and orientation of it only.

References

1. Gibson LJ, Ashby MF (1997) Cellular Solids: Structure and Properties, 2nd edn. Cambridge University Press, United Kingdom
2. Zhou Q (2001) Applications of cellular materials and structures in vehicle crashworthiness and occupant protection. In: Proceedings of the International Symposium on Plasticity and Impact. World Scientific Publishing, Zhuhai, China
3. Vian D, Bruneau M (2005) Steel plate shear walls for seismic design and retrofit of building structures. Multidisciplinary Center for Earthquake Engineering Research, University at Buffalo, State University of New York, Technical Report MCEER-05-0010

4. Keller D, Bruneau M (2008) Development of a steel plate shear wall bridge pier system conceived from a multi-hazard perspective. Multidisciplinary Center for Earthquake Engineering Research, University at Buffalo, State University of New York, Technical Report MCEER-08-0030
5. Chen L, Tsopelas P (2010) Cellular-solid shear walls under seismic excitations. In: Proceedings of 9th US National Conference on Earthquake Engineering, Toronto, 25–29 July 2010
6. Foti D, Diaferio M, Nobile R (2010) Optimal design of a new seismic passive protection device made in aluminium and steel. J Struct Eng Mech 35(1):119–122
7. Spadoni A (2008) Application of chiral cellular materials for the design of innovative components. Thesis for the for the Degree of Doctor of Philosophy, Georgia Institute of Technology
8. Papka SD, Kyriakides S (1994) In-plane compressive response and crushing of honeycomb. J Mech Phys Solids 42(10):1499–1532
9. Overaker DW, Cuitino AM, Langrana NA (1998) Effect of morphology and orientation on the behavior of two-dimensional hexagonal. Mech Mater 29:43–52
10. ABAQUS/Standard User's Manual, Version 6.14
11. Tsopelas P, Constantinou MC, Kircher CA et al (1997) Evaluation of Simplified Methods of Analysis for Yielding Structures. National Center for Earthquake Engineering Research, University at Buffalo, Technical Report NCEER-97-0012

Wind Response of a Bridge Pylon Using Numerical Simulations of the Atmospheric Boundary Layer and Fluid Structure Interaction (FSI)

Raúl Sánchez-García, Roberto Gomez, and J. Alberto Escobar

Abstract With the current computational capabilities and the development of Computational Fluid Dynamics (CFD), it is possible to perform numerical simulations of the neutral atmospheric boundary layer (ABL) to study the effects of wind forces on civil structures, and consequently evaluate their response with the purpose of obtaining a suitable design. In this article, we present numerical simulations of the fluid–structure interaction (FSI) of a single concrete pylon of a long span cable-stayed bridge to estimate its response under wind loads. In order to carry out a proper FSI modelling, it is necessary, firstly, to simulate the features of the atmospheric boundary layer (ABL) of the site where the structure will be located. This is used as input for the CFD model, which is studied with a special software providing pressure coefficients, velocity contours, and streamlines, which are then used to estimate wind loads acting on the structure. Secondly, a finite element analysis (FEA) is performed in order to evaluate the response of the structure in terms of mechanical elements (bending moment and shear) and displacements. In this paper, we applied this methodology to study the FSI response of a real bridge pylon. We compare results from the FEA analysis with those obtained from an isolated aero-elastic pylon scaled model studied in a wind tunnel.

1 Introduction

Winds disturbing civil structures and most of human activities are generated in the atmospheric boundary layer (ABL). Therefore, in order to estimate wind induced loads on buildings or bridges it is of main importance to have a reliable description of the ABL, which is not an easy task because wind is a turbulent phenomenon and its mathematical description is very complex [1]. Surface roughness of the ABL has a

R. Sánchez-García · R. Gomez (✉) · J. Alberto Escobar
Institute of Engineering, UNAM, Mexico City, Mexico
e-mail: RgomezM@iingen.unam.mx

R. Sánchez-García
e-mail: RsanchezG@iingen.unam.mx

J. Alberto Escobar
e-mail: JEscobarS@iingen.unam.mx

P. Gülkan et al. (eds.), *Developments in International Bridge Engineering*, Springer Tracts on Transportation and Traffic 17, https://doi.org/10.1007/978-3-030-59169-4_22

great influence in wind and turbulence patterns. To solve engineering problems such as predicting wind loads on civil structures, it is necessary to simulate the neutral atmospheric boundary layer in a boundary layer wind tunnel or perform numerical simulations with Computational Fluid Dynamics (CFD).

Regarding the behavior of cable-stayed bridges immersed in airflows, it is well known that they are highly susceptible to wind effects in both phases of construction and service, due to its high flexibility, low damping and geometry of the deck and pylons. Wind instability of pylons usually involves vertical bending and torsion modes; phenomena of vibration such as vortex shedding at low wind speeds or instability at high speeds can occur [2].

This paper presents the numerical study of a reinforced concrete pylon of a cable-stayed bridge including simulations of the ABL as well as the FSI response. Section 2 briefly describes the geometry of the bridge and in particular the pylon. Section 3 describes the study of the pylon in the wind tunnel. Section 4 briefly describes the fluid structure interaction phenomenon, as well as the computational model; finally, the conclusions of the study are presented in Sect. 5.

2 Description of the Structure

The "Baluarte" cable-stayed bridge is an icon of Mexican bridge engineering due to its characteristics and particular geographic location (Fig. 1). The bridge is part of an important highway in the North of Mexico. This area, apart from being very rugged topographically, is also prone to the passage of hurricanes as shown by observations recorded in the Pacific Ocean from 1949 until 2015 by the National Oceanic and Atmospheric Administration [3]; therefore, the bridge might be subjected to strong wind actions and consequently it is necessary to review its aerodynamic response under wind forces.

The bridge crosses a harsh topographic ravine with a depth of 390 m; it is the longest cable-stayed bridge built in North America with a total length of 1124 m including a main span of 520 m and two side sections of 354 and 250 m, respectively. A total of 152 cables in two vertical planes with a semi-harp array sustain the central main span. The overall dimensions of the bridge and pylon are shown in Fig. 2.

Fig. 1 Location and panoramic view of the "Baluarte" Bridge

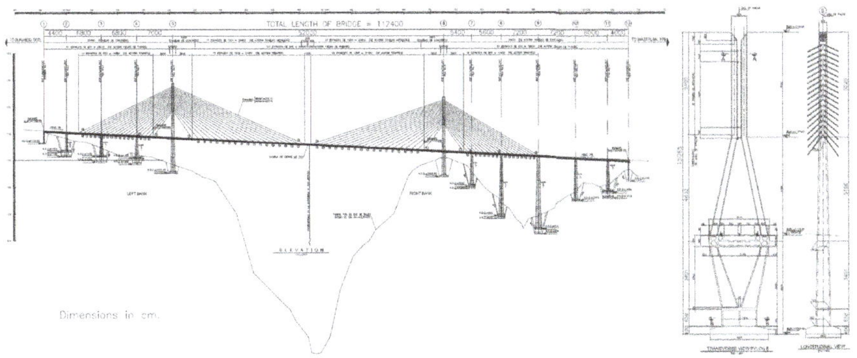

Fig. 2 Overall dimensions of the "Baluarte" Bridge and pylons

Fig. 3 Aero-elastic model of pylon 6. Wind tunnel test with an "open country" exposure type

Pylons located on axes 5 and 6 (Fig. 2), with heights of 169 and 152 m, respectively, are the most important elements of the substructure of the bridge. Pylons have a delta shape with hollow rectangular cross-sectional shape and variable wall thickness. These elements are slender and flexible which make them highly susceptible to wind effects.

3 Wind Tunnel Study of Pylon 6

An isolated (free-standing) aeroelastic model of pylon 6 was designed and built using a geometric scale of 1:250, and tested in a wind tunnel [4]. The response of the pylon was examined under two wind conditions: an atmospheric boundary layer flow with low turbulence and a typical flow of an open exposure terrain ("open country") [5]. The first condition was performed in order to identify the phenomena of vortex shedding and the second to have a more realistic fluid field. The pylon was tested in a wide range of speeds (from 10 to 90 m/s scaled, at the tip of the pylon) and wind incidence directions with increments of 10°. Figure 3 shows the pylon model

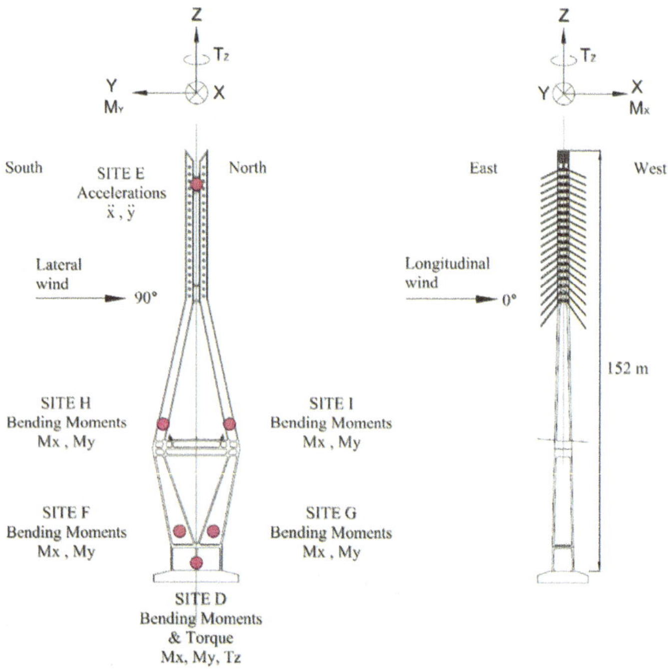

Fig. 4 Reference system and recording locations for the wind tunnel experimental tests

in the wind tunnel and Fig. 4 shows the points where the experimental response of the pylon was recorded.

4 Fluid–Structure Interaction, FSI

FSI models consider that solid surfaces act as interfaces between the fluid and the solid domains providing the means for transferring mechanical or thermal loads. CFD methods allow the solution of the fluid field, whereas the mechanics of solids and structures (finite element analysis, FEA) provides the state of stresses and deformations of the structure using the loads determined by a CFD analysis. Therefore, FSI analysis requires bidirectional interactions CFD-FEA (FSI-2Way) and a fully coupled iterative solution at each time step, that can be computationally expensive [6]. However, if the deformations of the structure are very small and time variations are also relatively small, the fluid field will barely be affected and the problem can be considered as one-way interaction (FSI-1Way). This approach is cheaper to apply and can be used to verify and validate the mathematical model, map loads on the surfaces, transfer mechanical or thermal loads and consequently provide a good estimate of the response. In this work, one-way or unidirectional FSI simulations are used to evaluate the response of the isolated pylon to wind loads.

4.1 Computational Model and Parameters

To determine wind loads, the characteristics of a neutral boundary layer was modeled for an "open country" type exposure (aerodynamic roughness of $z_0 = 0.03$ m [5]). Incident wind profiles were generated according to several reference speeds (10, 20, 30, 40, 50, 60, 70, 80 and 90 m/s) at a reference elevation equal to the height of the pylon (152 m). The mean wind profiles and turbulence quantities are imposed at the domain inlet according to Richards and Hoxey [7] and Blocken [8].

The CFD numerical model has an area of 140 × 250 m and 300 m in height with a nonstructured mesh type of 1.2 million of cells with a higher resolution near the walls of the pylon. The CFD commercial ANSYS - Fluent code was used in these simulations to solve the 3D RANS equations in steady state using the volume control method. For the closure problem, a realizable k-ε turbulence model was used.

On the other hand, ANSYS Mechanical package was used to generate the model of finite elements and calculate the structural response of the pylon to pressures caused by wind. Since the CFD model mesh is finer than the FEA mesh, aerodynamic loads at the surfaces of the pylon CFD model were transferred to the FEA model using an interpolation scheme.

4.2 FSI Results

With the FEA analysis and the specific wind loads derived from CFD analysis, flexure moments at sites F, G, H and I, for all the considered wind velocities were determined. Moments at the base of the pylon (site D) determined with simulations and those measured in the wind tunnel are presented in Fig. 5(a). It is observed that the numerical model provides about the same response as those determined in the wind tunnel. The same applies for sites F, G, H and I (Figs. 5(b) and 5(c)).

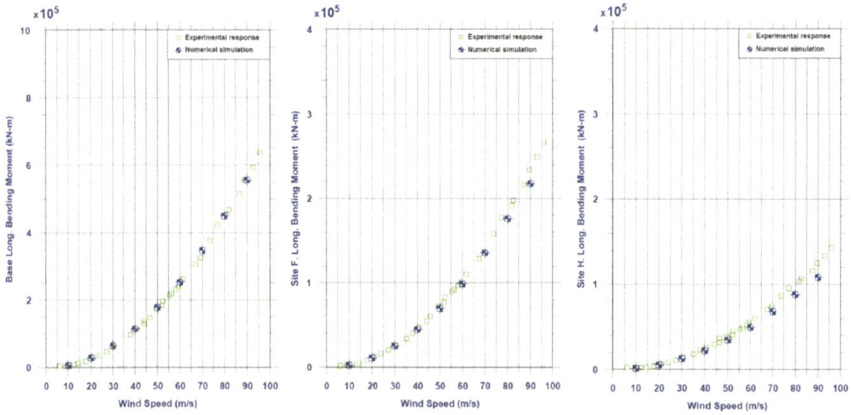

a). b). c).

Fig. 5 Pylon 6, FEA and CFD responses for the longitudinal bending moment. Sites D, F and H

5 Conclusions

CFD computational codes were used to replicate the characteristics of a neutral atmospheric boundary layer for a field corresponding to an "open country" type terrain. This was used to determine wind loads incident on the pylon of a cable-stayed bridge and study its structural response in terms of displacement and mechanical elements. The 3D behavior of the incident flow on the pylon and the pressure coefficients were used to corroborate some design hypothesis.

Results from one-way numerical simulations of fluid–structure interaction turned out to be approximate to those calculated in studies of the pylon in a wind tunnel. This fact corroborates the calibration of the mathematical model and demonstrates the feasibility of the application of the FSI in wind engineering problems. It is important to mention that the numerical simulations were performed at full scale and independently of the scale parameters used in the wind tunnel study.

References

1. Stull RB (1988) An Introduction to Boundary Layer Meteorology (1988 Edition). Kluwer Academic Publishers, Dordrecht, p 666
2. Strømmen EN (2006) Theory of Bridge Aerodynamics. Springer, Heidelberg
3. NOAA (2015). Historical Hurricane Tracks. https://www.coast.noaa.gov/hurricanes/#/app=3d30&db07-selectedIndex=0&3e3d-selectedIndex=0
4. King JPC, Kong L (2010) A study of wind effects for Baluarte Bridge México. London, Ontario, Canada
5. Wieringa J (1992) Updating the Davenport roughness classification. J Wind Eng Ind Aer 41–44:357–368

6. Sigrist JF (2015) Fluid-Structure Interaction: An Introduction to Finite Element Coupling, 1st edn. John Wiley & Sons Limited, Hoboken
7. Richards PJ, Hoxey RP (1993) Appropriate boundary conditions for computational wind engineering models using the k-ε turbulence model. J Wind Eng Ind Aerodyn 46 & 47:145–153
8. Blocken B, Stathopoulos T, Carmeliet J (2007) CFD simulation of the atmospheric boundary layer: wall function problems. Atmos Environ 41(2):238–252

The manufacturer's authorised representative in the EU is Springer
Nature Customer Service Centre GmbH, Europaplatz 3, 69115 Heidelberg,
Germany. If you have any concerns regarding our products, please
contact ProductSafety@springernature.com

Printed and bound by CPI Group (UK) Ltd, Croydon, CR0 4YY
28/09/2026
02224090-0011